KU-517-170

LEEDS BECKETT UNIVERSITY
LIBRARY

DISCARD

Leeds Metropolitan University

17 0098261 4

Principles of
Molecular Virology

For
Joseph

Principles of Molecular Virology

Alan J. Cann
University of Leicester, UK

ACADEMIC PRESS

Harcourt Brace & Company, Publishers
London   San Diego   New York
Boston   Sydney   Tokyo   Toronto

ACADEMIC PRESS LIMITED
24/28 Oval Road
LONDON NW1 7DX

United States Edition published by
ACADEMIC PRESS INC.
San Diego, CA 92101

Copyright © 1993, by
ACADEMIC PRESS LIMITED

This book is printed on acid-free paper

All Rights Reserved
No part of this book may be reproduced in any form by photostat,
microfilm, or any other means, without written permission
from the publishers

A catalogue record for this book is available from the British Library

ISBN: 0-12-158531-X✓

Cover photo courtesy of Institut Pasteur/CNRI/Science Photo Library

LEEDS METROPOLITAN
UNIVERSITY LIBRARY
1700982674
B4IIHV
109327 18 794
24 110 94
576.6 CAN

Typeset by Photo·graphics, Honiton, Devon
Printed in Great Britain by The University Printing House, Cambridge

Contents

Preface

This book came about through my own need for a text to augment my undergraduate lectures on virology. Not that there is a particular shortage of books on virology, but even during my own relatively short career, the subject has expanded so rapidly that most professionals feel pressurized by the task of keeping up with current trends, let alone the responsibility of initiating a new generation in such a rapidly changing discipline. Many excellent and recent texts exist that deal with the subject in a detailed and somewhat traditional manner; the second edition of *Fields'* *Virology* (Raven Press, 1990), the third edition of Matthews' *Plant Virology* (Academic Press, 1991) and, in a more general sense, the second edition of *Molecular Biology of the Cell* (Garland Press, 1989) immediately spring to mind. Unfortunately, these books have two major disadvantages for their use in teaching. First, the volume of material they contain is overwhelming for a student discovering the subject for the first time and who may be unable to sort out the information they require from the mountain of detail they encounter. Second, in these times, few students can afford to purchase one of these tomes. Even persuading university libraries, whose budgets are under constant pressure, to purchase more than one or two copies is difficult and an inadequate solution in a time of expanding student numbers. Better then to reserve these texts for the reference purpose for which they are best suited and to introduce students to the subject in a gentler way.

In discussions with many of my colleagues at this and other universities, it was clear that most felt there was a place for a text which would cover the current emphasis and concerns of virology. In these conversations, there was no doubt as to what was required. My contemporaries have no difficulty with the label 'molecular virology,' but most when pressed would have difficulty in expressing a comprehensive definition. Perhaps the best way to describe how I have approached the subject here is to consider it as 'virology at a molecular level' or even better, 'molecules and viruses.' Having already damned the 2000-page reference source as unsuitable for my purpose and yet set such an all-embracing definition of 'molecular virology,' the problem was therefore how to resolve these two apparently conflicting issues. My chosen solution is to outline the *principles* of the subject with reference to specific *examples* chosen to illustrate the matter under discussion. The onus is therefore firmly on the reader to pursue particular matters on which he or she requires more information in more detailed 'reference texts' or in the immense volume of research publications appearing annually.

I would like to have spent much more time discussing the history of virology in Chapter 1, a subject I find to be a fascinating as well as a valuable insight as to how we got to where we are today. In the event, it was only possible to provide a brief overview and to refer the reader to one of the many other texts which have been published on this subject – perhaps the one area of virology where an author's work does not become outdated in a short period. It is only my intention to arm readers with the framework that makes it possible for them to achieve this task successfully. Anyone who complains that this book does not spend sufficient time dealing with (or even mention) their pet area of interest has therefore missed the point.

To disarm the jargon (which I have avoided wherever possible) and unavoidable technical terms, I have included a glossary as an appendix to the book. Terms shown in the text in **bold** print are defined in this glossary.

Cliché it may be, but there are genuinely too many people to acknowledge for the creation of this book to make it possible to do this individually. It will have to suffice for me to thank all my colleagues in Leicester and elsewhere for helpful discussions; all the people who have helped and influenced my career over the years, and the undergraduates on whom I have field-tested the material, and for whom this book is intended.

Alan J. Cann
Microbiology Department
University of Leicester
October 1992

Chapter 1

Introduction

'Those who cannot remember the past are condemned to repeat it'.
(George Santayana)

There is more biological diversity within viruses than in all the rest of the bacterial, plant and animal kingdoms put together. This is the result of the success of viruses in parasitizing all known groups of living organisms, and understanding this diversity is the key to comprehending the interactions of viruses with their hosts. This book deals with 'molecular virology' in a rather broad sense – that is, 'virology at a molecular level,' or perhaps even, 'molecules and viruses.' Protein–protein, protein–nucleic acid, and protein–lipid interactions determine the structure of virus particles, the synthesis and expression of virus genomes and the effects of viruses on the host cell. This is virology at a molecular level.

However, before exploring the subject further, it is necessary to understand the nature of viruses. It would also be useful to know something of the history of virology, or more accurately, how virology as a discipline in its own right arose, in order to understand its current concerns and future directions. These are the purposes of this introductory chapter. The principles behind certain techniques mentioned in this chapter may not be familiar to some readers. It may be helpful to use the references at the end of this chapter to become conversant with these methods. In this and the subsequent chapters, terms in the text in **bold** print are defined in the glossary in Appendix 2.

Viruses are Distinct from 'Living' Organisms

Viruses are submicroscopic, obligate intracellular parasites. This simple but useful definition goes a long way towards describing and differentiating viruses from all other groups of living organisms. However, this short definition is in itself inadequate. Clearly, there is no problem in delimiting viruses from higher 'macroscopic' organisms. Even within a broad definition of microbiology encompassing prokaryotic organisms and microscopic eukaryotes such as algae, protozoa and fungi, in most cases it will suffice. There are, however, a few groups of prokaryotic organisms which have specialized parasitic life-cycles and which confound the above definition.

These are the Rickettsiae and Chlamydiae, obligate intracellular parasitic bacteria which have evolved to be so cell-associated that they can only exist outside the cells of their hosts for a short period of time before losing their infectivity. Therefore it is necessary to add further clauses to the definition of what constitutes a virus:

- Virus particles are produced from the assembly of pre-formed components, whereas other agents 'grow' from an increase in the integrated sum of their components and reproduce by division.
- Virus particles (**virions**) themselves do not 'grow' or undergo division.
- Viruses lack the genetic information which encodes apparatus necessary for the generation of metabolic energy or for protein synthesis (ribosomes).

No known virus has the biochemical or genetic potential to generate the energy necessary for driving all biological processes (e.g. macromolecular synthesis). They are therefore absolutely dependent on the host cell for this function. It is often asked whether viruses are 'alive' or not. One view is that inside the host cell, viruses are 'alive,' whereas outside it they are merely complex assemblages of metabolically inert chemicals. That is not to say that chemical changes do not occur in extracellular virus particles, as will be explained elsewhere, but these are in no sense the 'growth' of a living organism.

Although there will always be some exceptions and uncertainties in the case of organisms which are too small to see and in many cases very difficult to study, for the most part, the above guidelines will suffice to determine the definition of a virus. Unfortunately, molecular investigation of certain infectious agents has revealed further complications. A number of novel, pathogenic entities possess properties which confound the above definition, yet are clearly more similar to viruses than other organisms. These are the entities known as **viroids**, **virusoids**, and **prions**. Viroids are very small (200–400 nucleotides) circular RNA molecules with a rod-like secondary structure. They have no capsid or envelope and are associated with certain plant diseases. Their replication strategy is like that of viruses – they are obligate intracellular parasites. Virusoids are **satellite**, viroid-like molecules, somewhat larger than viroids (e.g. approximately 1000 nucleotides), which are dependent on the presence of virus replication for multiplication (hence 'satellite'); they are packaged into virus capsids as passengers. Prions are rather ill-defined infectious agents consisting of a single type of protein molecule with no nucleic acid component. Confusion arises from the fact that the prion protein and the gene that encodes it are also found in normal 'uninfected' cells. These agents are associated with 'slow virus diseases' such as Creutzfeldt–Jakob disease in humans, scrapie in sheep and bovine spongiform encephalopathy (BSE) in cattle. Chapter 8 deals with these novel infectious agents in more detail. Moreover, molecular genetics has shown that perhaps 5–10% of the eukaryotic cell genome is composed of mobile retrovirus-like elements (**retrotransposons**), which may have had a considerable role in shaping these complex genomes (Chapter 3). Furthermore, certain **bacteriophage** genomes closely resemble bacterial plasmids in their structure and in the way they are replicated. Therefore, recent research has revealed that the relationship between viruses and other living organisms is perhaps more complex than was previously thought.

The Origins of Virology

It is easy to regard events which occurred prior to our own personal experience as pre-historic. Thus much has been written about virology as a 'new' discipline in biology and as far as the formal recognition of viruses as distinct from other living organisms is concerned, this is true. However, we now realize that not only were ancient peoples aware of the effects of virus infection, but in some instances also carried out active research into the causes and prevention of virus diseases. Perhaps the first written record of a virus infection consists of a hieroglyph from Memphis, the capital of ancient Egypt, drawn in approximately 1400BC, which depicts a temple priest showing typical clinical signs of paralytic poliomyelitis. In addition, the Pharoh Ramses V, who died in 1196BC and whose extraordinarily well-preserved mummified body is now in Cairo Museum, is believed to have succumbed to smallpox – the comparison between the pustular lesions on the face of the mummy and those of more recent patients is startling.

Smallpox was endemic in China by 1000BC. In response, the practice of **variolation** was developed. Recognizing that survivors of smallpox outbreaks were protected from subsequent infection, the Chinese inhaled the dried crusts from smallpox lesions like snuff, or in later modifications, inoculated the pus from a lesion into a scratch on the forearm of a child. This variolation was practiced for centuries and was shown to be an effective, although risky, method of disease prevention, since the outcome of the inoculation was never certain. Edward Jenner was nearly killed by variolation at the age of seven! Not surprisingly, this experience spurred him on to find a safer alternative treatment, and on 14 May 1796, he used cowpox-infected material obtained from the hand of Sarah Nemes, a milkmaid from his home village of Berkeley in Gloucestershire, England, to successfully vaccinate 8-year-old James Phipps. Although initially controversial, **vaccination** against smallpox was almost universally adopted worldwide during the nineteenth century.

This early success, although a triumph of scientific observation and reasoning, was not based on any real understanding of the nature of infectious agents, which arose separately from another line of reasoning. Antony van Leeuwenhoek (1632–1723), a Dutch merchant, constructed the first simple microscopes and with these, identified bacteria as the 'animalcules' he saw in his specimens. However, it was not until Robert Koch and Louis Pasteur in the 1880s jointly propounded the 'germ theory' of disease that the significance of these organisms became apparent. Koch defined the four famous criteria now known as Koch's postulates which are still generally regarded as the proof that an infectious agent is responsible for a specific disease:

(1) The agent must be present in every case of the disease.
(2) The agent must be isolated from the host and grown *in vitro*.
(3) The disease must be reproduced when a pure culture of the agent is inoculated into a healthy susceptible host.
(4) The same agent must be recovered once again from the experimentally infected host.

Subsequently, Pasteur worked extensively on rabies, which he identified as being caused by a 'virus' (from the Latin for 'poison') but in spite of this, he did not discriminate between bacteria and other agents of disease. In 1892, Dimitri Iwanowski, a Russian botanist, showed that extracts from diseased tobacco plants could transmit disease to other plants after passage through ceramic filters fine enough to retain the smallest known bacteria. Unfortunately, he did not realize the full significance of these results. A few years later, in 1898, Martinus Beijerinick confirmed and extended Iwanowski's results on tobacco mosaic virus and was the first to develop the modern idea of the virus, which he referred to as *contagium vivum fluidum* ('soluble living germ'). Freidrich Loeffler and Paul Frosch (1898) showed that a similar agent was responsible for foot-and-mouth disease in cattle, but despite the realization that these new-found agents caused disease in animals as well as plants, there was resistance to the idea that they might have anything to do with human diseases. This was finally dispelled by Karl Landsteiner and E. Popper in 1909, who showed that poliomyelitis was caused by a 'filterable agent' – the first human disease to be recognized as having a viral cause.

Frederick Twort (1915) and Felix d'Herelle (1917) were the first to recognize viruses which infect bacteria, which d'Herelle called **bacteriophages** (eaters of bacteria). In the 1930s and subsequent decades, pioneering virologists such as S.E. Luria, M. Delbrück, and many others utilized these viruses as model systems to investigate many aspects of virology, including virus structure (Chapter 2), genetics (Chapter 3), replication (Chapter 4), etc. These relatively simple agents have since been very important in the development of our understanding of all types of viruses, including those of humans which are much more difficult to propagate and study. The further history of virology is the story of the development of experimental tools and systems with which viruses could be examined and by which whole new areas of biology were opened up – including not only the biology of the viruses themselves, but inevitably also the biology of the host cells on which these agents are entirely dependent.

Living host systems

In 1881, Louis Pasteur began to study rabies in animals. Over a number of years, he developed methods of producing **attenuated** virus preparations by progressively drying the spinal cords of rabbits experimentally infected with the agent which, when inoculated into animals, would protect from challenge with virulent virus. This was the first artificially produced virus **vaccine**, since the ancient practice of variolation (above) and Jenner's use of cowpox virus for vaccination relied on naturally occurring viruses. Whole plants have been used to study the effects of plant viruses after infection ever since tobacco mosaic virus was first discovered by Iwanowski, usually by rubbing preparations containing virus particles into the leaves or stem of the plant.

During the Spanish–American War of the late nineteenth century and the subsequent building of the Panama Canal, American deaths due to yellow fever were colossal. The disease also appeared to be spreading slowly northward into the continental United States. Through experimental transmission to mice, in 1900 Walter Reed demonstrated that yellow fever was caused by a virus, spread by mosquitoes. This discovery eventually enabled Max Theiler in 1937 to propagate the virus in chick embryos and to produce an attenuated vaccine – the 17D strain – which is still in use today.

The success of this approach led many other investigators from the 1930s to the 1950s to develop animal systems to identify and propagate pathogenic viruses. **Eukaryotic** cells can be grown *in vitro* ('tissue culture') and viruses can be propagated in these cultures, but these techniques are expensive and technically quite demanding. Some viruses will replicate in the living tissues of developing embryonated hens eggs, such as influenzavirus. Egg-adapted strains of influenzavirus replicate well in eggs and very high virus **titers** can be obtained. Embryonated hens eggs were first used to propagate viruses in the early decades of this century. This method has proved to be highly effective for the isolation and culture of many viruses, particularly strains of influenzavirus and various poxviruses (e.g. vaccinia virus). Counting the 'pocks' on the chorioallantoic membrane of eggs produced by the replication of vaccinia virus was the first quantitative assay for any virus. Animal host systems still have their uses in virology:

- to produce viruses which cannot be propagated *in vitro*, e.g. hepatitis B virus;
- to study the pathogenesis of virus infections, e.g. coxsackieviruses; and
- to test vaccine safety, e.g. oral poliovirus vaccine.

Nevertheless, they are increasingly being discarded for the following reasons:

- Breeding and maintenance of animals infected with pathogenic viruses is expensive.
- Whole animals are complex systems, in which it is sometimes difficult to discern events.
- Results obtained are not always reproducible, due to host variation.
- Unnecessary or wasteful use of experimental animals is morally repugnant.
- They are rapidly being overtaken by 'modern science' – cell culture and molecular biology.

The use of whole plants as host organisms does not engender the same moral objections as the use of living animals and continues to play an important part in the study of plant viruses, although such systems are sometimes slow to deliver results and expensive to maintain.

In recent years, an entirely new technology has been employed to study the effects of viruses on host organisms. This involves the creation of **transgenic** animals and plants by inserting all or part of the virus genome into the DNA of the experimental organism, resulting in expression of virus mRNA and proteins in somatic cells (and sometimes in the cells of the germ line). Thus the pathogenic effects of virus proteins, individually and in various combinations, can be studied in living hosts. While this technique often raises the same moral objections as 'old-fashioned' experimental infection of animals by viruses, it is clearly an immensely powerful new tool for the study of viral pathogenicity. A growing number of plant and animal viruses genes have been analyzed in this way, but the results have not always been as expected, and in many cases it has proved difficult to equate the observations obtained with those gathered from experimental infections. Nevertheless, this method will undoubtedly become much more widely used as more of the technical difficulties associated with the construction of transgenic organisms are solved.

Cell culture methods

Cell culture began early this century with whole organ cultures, then progressed to methods involving individual cells, either **primary cell** cultures (somatic cells from an experimental animal or taken from a human patient which can be maintained for a short period in culture) or **immortalized cell** lines, which, given appropriate conditions, continue to grow in culture indefinitely.

In 1949, J.F. Enders and his colleagues were able to propagate poliovirus in primary human cell cultures. This opened what many regard as the 'Golden Age of Virology' – the identification and isolation during the 1950s and 1960s of many viruses and their association with human diseases, for example, many enteroviruses and respiratory viruses, such as adenoviruses. Widespread virus isolation led to the realization that subclinical virus infections were very common, for example, even in epidemics of the most virulent strains of poliovirus, there are approximately 100 inapparent infections for each paralytic case of poliomyelitis.

Renato Dulbecco in 1952 was the first to quantify accurately animal viruses using a plaque assay. In this technique, dilutions of the virus are used to infect a cultured cell **monolayer**, which is then covered with soft agar to restrict diffusion of the virus, resulting in localized cell killing and the appearance of **plaques** after the monolayer is stained (Figure 1.1). Counting the number of plaques directly determines the number of infectious virus particles applied to the plate. The same technique can also be used biologically to clone a virus, i.e. isolate a pure form from a mixture of types. This technique had been in use for some time to quantify the number of infectious virus particles in bacteriophage suspensions applied to confluent 'lawns' of bacterial cells on agar plates, but its application to eukaryotic viruses enabled rapid advances in the study of virus replication to be made. Plaque assays largely replaced earlier endpoint dilution techniques, such as the $TCID_{50}$ (tissue culture infectious dose) assay, which represent statistical means of measuring virus populations in culture. However, endpoint techniques may still be used in certain circumstances, for example, for viruses which do not replicate in culture, or which are not cytopathic and do not produce plaques (e.g. human immunodeficiency virus (HIV)).

Serological/immunological methods

As virology was emerging in the early part of this century, the techniques of immunology were also developing and, as with molecular biology in more recent years, the two disciplines have always been very closely linked. Understanding mechanisms of immunity to virus infections has, of course, been very important. Recently, the role that the immune system itself plays in pathogenesis has become known (see Chapter 7). Immunology as a discipline in its own right has contributed many of the classical techniques to virology (Figure 1.2).

George Hirst in 1941 observed **hemagglutination** of red blood cells by influenza virus (see Chapter 4). This proved to be an important tool not only in

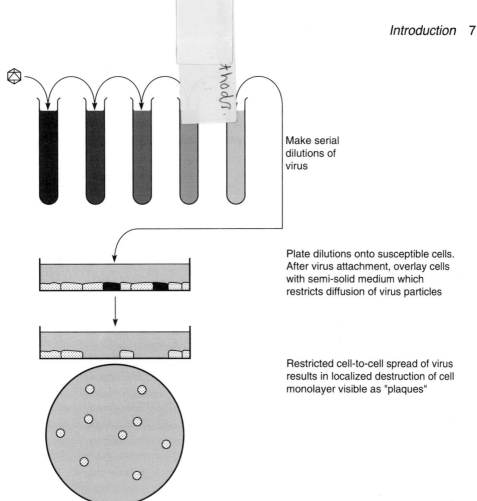

Make serial
dilutions of
virus

Plate dilutions onto susceptible cells.
After virus attachment, overlay cells
with semi-solid medium which
restricts diffusion of virus particles

Restricted cell-to-cell spread of virus
results in localized destruction of cell
monolayer visible as "plaques"

Figure 1.1 Plaque assays are performed by applying a suitable dilution of a virus preparation to a confluent or semi-confluent adherent monolayer of susceptible cells. After allowing time for virus attachment to and infection of the cells, liquid medium is replaced by a semi-solid culture medium containing a polymer such as agarose or carboxymethyl cellulose, which restricts diffusion of virus particles from infected cells. Only direct cell-to-cell spread can occur, resulting in localized destruction of the monolayer. After a suitable period, the medium is usually removed and the cells stained to make the holes in the monolayer (**plaques**) more easily visible. Each plaque therefore results from infection by a single **plaque-forming unit**.

the study of influenza, but also with several other groups of viruses, for example, rubella virus. In addition to measuring the **titer** (i.e. relative amount) of virus present in any preparation, this technique can also be used to determine the antigenic type of the virus. Hemagglutination will not occur in the presence of antibodies which bind to and block the virus hemagglutinin. If an antiserum is titrated against a given number of hemagglutinating units, the hemagglutination inhibition titer and specificity of the antiserum can be determined. Also, if antisera of known specificity are used to inhibit hemagglutination, the antigenic type of an

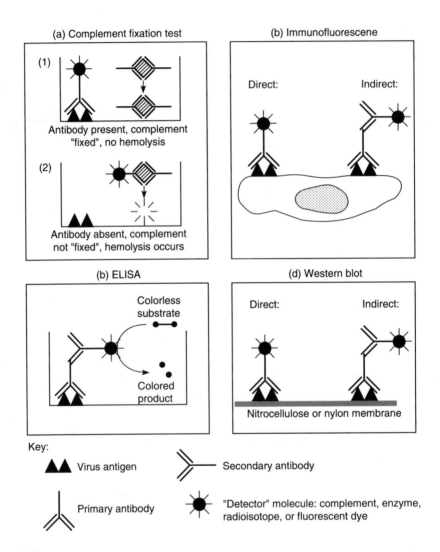

Figure 1.2

unknown virus can be determined. In the 1960s and subsequent years, many improved detection methods for viruses were developed, for example:

- Complement fixation tests
- Radioimmunoassays
- Immunofluorescence (direct detection of virus antigens in infected cells or tissue)
- Enzyme-linked immunosorbent assays (ELISAs)
- Radioimmune precipitation
- Western blot assays

These techniques are sensitive, quick and quantitative.

In 1975, Kohler and Milstein isolated the first monoclonal antibodies from clones of cells selected *in vitro* to produce an antibody of a single specificity directed against a particular antigenic target. This enabled virologists to look not only at the whole virus, but at specific regions – epitopes – of individual virus antigens

---

Figure 1.2 It is hard to overestimate the importance of serological techniques in virology. Four assays illustrated in this figure have been used for many years and are of widespread value. (a) The *complement fixation test* works on the basis that complement is sequestered by antigen–antibody complexes. 'Sensitized' antibody-coated red blood cells and known amounts of complement plus a particular virus antigen are added to the wells of a multi-well plate, plus the serum to be tested. In the absence of antibodies directed against the virus antigen, some free complement remains and this causes the lysis of the sensitized red blood cells (hemolysis). If, however, the test serum contains a sufficiently high titre of antiviral antibodies, no free complement remains and hemolysis does not occur. Titrating the test serum by means of serial dilutions allows a quantitative measurement of the amount of antivirus antibody present to be made. (b) *Immunofluorescence* is performed using derivatized antibodies containing a covalently-linked fluorescent molecule which emits a characteristically colored light when illuminated by light of a different wavelenth,e.g. rhodamine (red) or fluorescein (green). In *direct immunofluorescence*, the antivirus antibody itself is conjugated to the fluorescent marker, whereas in *indirect immunofluorescence*, a second antibody reactive to the antivirus antibody carries the marker. Immunofluorescence can be used not only to identify virus-infected cells in populations of cells or in tissue sections, but also to determine the subcellular localization of particular virus proteins, e.g. in the nucleus or in the cytoplasm. (c) *Enzyme-linked immunosorbent assays (ELISA)* are a rapid and sensitive means of identifying or quantifying small amounts of virus antigens or antivirus antibodies. Either an antigen (in the case of an ELISA to detect antibodies) or antibody (in the case of an antigen ELISA) are bound to the surface of a multi-well plate. An antibody specific for the test antigen which has been conjugated with an enzyme molecule (such as alkaline phosphate or horseradish peroxidase) is then added. As with immunofluorescence, ELISA assays may rely on direct or indirect detection of the test antigen. During a short incubation, a colorless substrate for the enzyme is converted to a colored product, thus amplifying the signal produced by a very small amount of antigen. The intensity of the product can easily be measured in a specialized spectrophotometer (a 'plate reader'). ELISA assays can be mechanized and are therefore suitable for routine tests on large numbers of clinical samples. (d) *Western blotting* is used to analyze a specific virus protein from a complex mixture of antigens. Virus antigen-containing preparations (particles, infected cells, or clinical materials) are subjected to electrophoresis on a polyacrylamide gel. Proteins from the gel are then transferred to a nitrocellulose or nylon membrane and immobilized in their relative positions from the gel. Specific antigens are detected by allowing the membrane to react with antibodies directed against the antigen of interest. By the use of samples containing proteins of known sizes in known amounts, the apparent molecular weight and relative amounts of antigen in the test samples can be determined.

(Figure 1.3). In recent years, this ability has greatly increased our understanding of the function of individual virus proteins. Monoclonal antibodies are also finding increasingly widespread application in other types of serological assay (e.g. ELISAs), to increase their reproducibility, sensitivity, and specificity.

It would be inappropriate here to spend too long on the technical details of what is also a very rapidly expanding field of knowledge. However, I recommend strongly that readers who are not familiar with the techniques mentioned above should familiarize themselves thoroughly with this subject by reading one or more of the texts given in the references for this chapter. Time spent in this way will repay the reader throughout the rest of this book.

Immunize animals with antigen (complex mixture of epitopes)

Spleen cells

Immortal B-cell line

CELL FUSION

Apply selective medium to fusions, grow on hybridomas

Test supernatant for antibody specificity

Biologically clone cells by limiting dilution

Grow up antibody secreting cells and isolate antibody from medium

Figure 1.3 Monoclonal antibodies are produced by immunization of an animal with an antigen which usually contains a complex mixture of epitopes. Immature B-cells are later prepared from the spleen of the animal and these are fused with a myeloma cell line, resulting in the formation of **transformed** cells continuously secreting antibodies. A small proportion of these will make a single type of antibody (a monoclonal antibody) against the desired epitope. Recently, *in vitro* molecular techniques have been developed to speed up the selection of monoclonal antibodies, although these have not yet replaced the original approach shown here.

Ultrastructural studies

These can be considered under three areas: physical methods, chemical methods, and electron microscopy. Physical measurements of virus particles began with the earliest determinations of their size by filtration through colloidal membranes of various pore sizes in the 1930s. Experiments of this sort led to the first (rather inaccurate) estimates of the size of virus particles. The accuracy of these was improved greatly by studies of the sedimentation properties of viruses in ultracentrifuges in the 1960s (Figure 1.4). Differential centrifugation proved to be of great use in obtaining purified and highly concentrated preparations of many

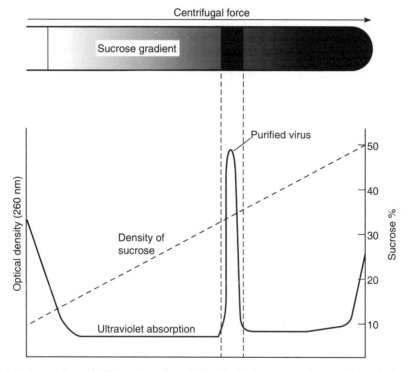

Figure 1.4 A number of different sedimentation techniques can be used to study viruses. In *rate zonal centrifugation* (shown here), virus particles are applied to the top of a pre-formed density gradient, i.e. a sucrose or salt solution of increasing density from the top to the bottom of the tube (top of figure). After a period of time in an ultracentrifuge, the gradient is separated into a number of fractions, which are analyzed for the presence of virus particles. In the figure, the nucleic acid of the virus genome is detected by its absorption of ultraviolet light (below). This method can be used both to purify virus particles or nucleic acids and to determine their sedimentation characteristics. In *equilibrium* or *isopycnic centrifugation*, the sample is present in a homologous mixture containing a dense salt such as cesium chloride. A density gradient forms in the tube during centrifugation and the sample forms a band at a position in the tube equivalent to its own density. This method can thus be used to determine the density of virus particles and is commonly used to purify plasmid DNA.

different viruses, free of contamination from host cell components which can be subjected to chemical analysis. The relative density of virus particles, measured in solutions of sucrose or CsCl is also a characteristic feature, revealing information about the proportions of nucleic acid and proteins in the particles.

The physical properties of viruses can be determined by spectroscopy, using either ultraviolet light to examine the nucleic acid content of the particle or visible light to determine its light-scattering properties. Electrophoresis of intact virus particles has yielded some limited information, but electrophoretic analysis of individual virion proteins by gel electrophoresis, and particularly also of nucleic acid genomes (Chapter 3), has been far more valuable. However, by far the most important method for the elucidation of virus structures has been the study of X-ray (or other types of electromagnetic radiation) diffraction by crystalline forms of purified virus. This technique permits determination of the structure of virions at an atomic level.

The complete structures of many viruses, representative of many of the major groups, have now been determined at a resolution of a few angstroms (see Chapter 2). This has advanced our understanding of the functions of the viral particle considerably. However, a number of viruses have proved to be resistant to this type of investigation, a fact which highlights some of the problems inherent in this otherwise powerful technique. For a start, the virus must first be purified to quite a high degree, otherwise specific information on the virus cannot be gathered. This presupposes that adequate quantities of the virus can be propagated in culture or obtained from infected tissues or patients and that a method is available to purify virus particles without loss of structural integrity. In a number of important cases, this requirement rules out further study (e.g. hepatitis B virus). Second, the purified virus must be able to form paracrystalline arrays big enough to cause significant diffraction of the radiation source. For some viruses, this is relatively straightforward and crystals big enough to see with the naked eye and which diffract strongly are easily formed. This is particularly true for a number of plant viruses, such as tobacco mosaic virus (TMV) (which was first crystallized by W.M. Stanley in 1935) and turnip yellow mosaic virus (TYMV), the structures of which were among the first to be determined during the 1950s. It is significant that these two viruses represent the two fundamental types of virus particle: helical in the case of TMV and isometric for TYMV (see Chapter 2). In many cases, however, only microscopic crystals can be prepared. A partial answer to this problem is to use ever more powerful radiation sources which enable good data to be collected from small crystals. Powerful synchotron sources which generate intense beams of radiation have been built during the last few decades and are now used extensively for this purpose. However, there is a limit beyond which this brute force approach fails to yield further benefit. A number of important viruses steadfastly refuse to crystallize – this is a particularly common problem with irregularly shaped viruses, for example those which have an outer lipid envelope, and to date, no complete high-resolution atomic structure has yet been determined for many viruses of this type (e.g. HIV). Modifications of the basic diffraction technique (such as electron scattering by membrane-associated protein arrays and cryo-electron microscopy) may help to provide more information in the future, but it is unlikely that these variations will solve this problem completely. One further limitation is that some of the largest virus particles, such as poxviruses, contain hundreds of different proteins and are at present too complex to be analyzed using these techniques.

Nuclear magnetic resonance (NMR) is increasingly being used to determine the atomic structure of all kinds of molecules, including proteins and nucleic acids. The limitation of this method is that only relatively small molecules can be analyzed before the signals obtained become so confusing that they are impossible to decipher with present technology. At present, the upper size limit for this technique restricts its use to molecules with a molecular weight of less than about 30 000–40 000 – considerably less than even the smallest virus particles. Nevertheless, this method may well prove to be of value in the future, certainly for examining isolated viral proteins, if not for intact virions.

Chemical investigation can be used to determine not only the overall composition of viruses and the nature of the nucleic acid which comprises the virus genome, but also the construction of the particle and the way in which individual components relate to each other in the capsid. Many classic studies of virus structure have been based on gradual, stepwise disruption of particles by slow alteration of pH, or gradual addition of protein-denaturing agents such as urea, phenol, or detergents. Under these conditions, valuable information can sometimes be obtained from relatively simple experiments. For example, as urea is gradually added to preparations of purified adenovirus particles, they break down in an ordered, stepwise fashion which releases subviral protein assemblies, revealing how the particles are made up. In the case of TMV, similar studies of capsid organization have been performed by renaturation of the capsid protein under various conditions (Figure 1.5). In simple terms, the reagents used to denature virus capsids can indicate the basis of

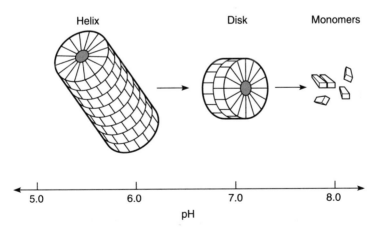

Figure 1.5 The structure and stability of virus particles can be examined by progressive denaturation or renaturation studies. At any particular ionic strength, the purified capsid protein of tobacco mosaic virus (TMV) spontaneously assembles into different structures, dependent on the pH of the solution. At a pH of around 6.0, the particles formed have a helical structure very similar to infectious virus particles. As the pH is increased to about 7.0, disk-like structures are formed. At even higher pH values, individual capsid monomers fail to assemble into more complex structures.

the stable interactions between its components. Proteins bound together by electrostatic interactions can be eluted by addition of ionic salts or alteration of pH, those bound by non-ionic, hydrophobic interactions can be eluted by reagents such as urea, and proteins which interact with lipid components can be eluted by non-ionic detergents or organic solvents.

In addition to its fundamental structure, progressive denaturation can be used to observe alteration or loss of antigenic sites on the surface of particles and in this way, a picture of the physical state of the particle can be built up. Proteins exposed on the surface of viruses can be labeled with various compounds (e.g. iodine) to indicate which parts of the protein are exposed and which are protected inside the particle or by lipid membranes. Cross-linking reagents such as psoralens, or newer synthetic reagents with side-arms of specific lengths are used to determine the spatial relationship of proteins and nucleic acids in intact viruses.

Electron microscopes, developed in the 1940s, overcome the fundamental limitation of light microscopes from the virologist's point of view, i.e. the inability to resolve individual virus particles due to physical constraints caused by the wavelength of visible light illumination and the optics of the instruments. During subsequent years, techniques were developed which allowed the direct examination of viruses at magnifications of over $100\,000\times$. There are two fundamental types of electron microscope, the transmission electron microscope (TEM) and the scanning electron microscope (SEM) (Figure 1.6). Although beautiful images with the appearance of three dimensions are produced by the SEM, for practical investigations of virus structure, the higher magnifications achievable with the TEM have proved to be of most value. Two fundamental types of information can be obtained by electron microscopy of viruses: the absolute number of virus particles present in any preparation (total count) and the appearance and structure of the virions (see below). Electron microscopy can provide a rapid method of virus detection and diagnosis, but in itself, may give misleading information. Many cellular components, for example ribosomes, can resemble 'virus-like particles,' particularly in crude preparations. This difficulty can be overcome by the use of antisera specific for particular virus antigens conjugated to electron-dense markers such as the iron-containing protein ferretin, or colloidal gold suspensions. This highly specific technique, known as immunoelectron microscopy, is gaining ground as a rapid method for diagnosis.

In recent years, developments in electron microscopy have allowed investigation of the structure of fragile viruses which cannot be determined by X-ray crystallography. These include cryo-electron microscopy, in which the virus particles are maintained at very low temperatures on cooled specimen stages; examination of particles embedded in vitreous ice, which does not disrupt the particles by the formation of ice crystals; low-irradiation electron microscopy, reducing the destructive bombardment of the specimen with electrons; and sophisticated image-analysis and image-reconstruction techniques which permit accurate, three-dimensional images to be formed from multiple images that individually would appear as very poor quality.

In the late 1950s, Brenner and Horne (among others) developed sophisticated techniques which enabled them to use electron microscopy to reveal many of the fine details of the structure of virus particles. One of the most valuable techniques proved to be the use of electron-dense dyes such as phosphotungstic acid or uranyl acetate to examine virus particles by negative staining. The small metal ions in such dyes are able to penetrate the minute crevices between the protein subunits in a viral capsid to reveal the fine structure of the particle. Using such data, Francis Crick and James Watson (1956) were the first to suggest that virus capsids are composed of numerous identical protein subunits arranged either in helical or cubic (icosahedral) symmetry. In 1962, Caspar and Klug extended these observations and elucidated the fundamental principles of symmetry which allow repeated protomers to form virus capsids, based on the principle of **quasi-equivalence** (see Chapter 2). This combined theoretical and practical approach has resulted in our present understanding of the structure of viruses of many different types.

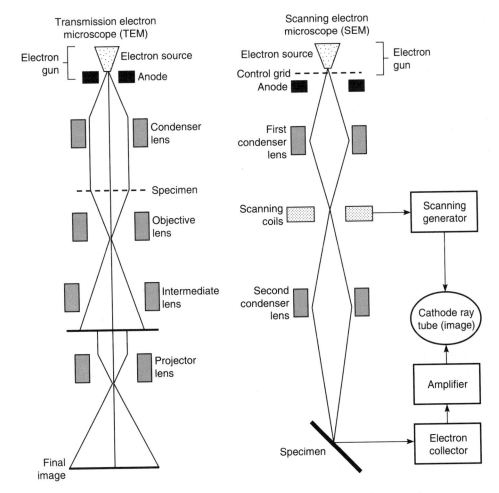

Figure 1.6 Working principles of transmission and scanning electron microscopes.

'Molecular Biology'

All the above techniques of investigation are themselves 'molecular biology' in the original sense of the term. However, during the last two decades the term 'molecular biology' ('genetic engineering' or 'genetic manipulation') has taken on a new and different meaning. These new techniques for manipulating nucleic acids *in vitro* (that is, outside of living cells or organisms) do not comprise a new discipline, but are an outgrowth of earlier developments in biochemistry and cell biology over the previous 50 years. Despite this, this powerful new technology has revolutionized virology and, to a large extent, has shifted the focus of attention away from the virus particle onto the virus genome. Once again, this book is not the place to discuss in detail the technical aspects of these methods, and readers are referred to one of the many texts published in recent years, such as those given at the end of this chapter.

Virus infection has long been used to probe the working of 'normal' (i.e. uninfected) cells, for example, to look at macromolecular synthesis. This is true, for example, of the applications of bacteriophages in bacterial genetics, and also in many instances where the study of eukaryotic viruses has revealed fundamental information about the cell biology and genomic organization of higher organisms. In 1970, J. Kates first observed that vaccinia virus mRNAs were polyadenylated at their 3′ ends and in the same year, Howard Temin and David Baltimore jointly identified the enzyme reverse transcriptase (RNA-dependent DNA polymerase) in retrovirus-infected cells. This shattered the so-called 'central dogma' of biology, i.e. that there is a one-way flow of information from DNA through RNA into protein, and revealed the plasticity of the eukaryotic genome. Subsequently, the purification of this enzyme from retrovirus particles permitted cDNA cloning, which greatly accelerated the study of viruses with RNA genomes – a good illustration of the catalytic nature of scientific advances. In 1977, Sharp recognized that adenovirus mRNAs were spliced to remove intervening sequences, indicating the similarities between viral and cellular genomes.

Initially at least, the effect of this new technology was to shift the emphasis of investigation from proteins to nucleic acids. As the power of the techniques developed, it quickly became possible to determine the nucleotide sequences of entire virus genomes, beginning with the smallest bacteriophages in the mid-1970s and working up to the largest of all virus genomes, those of the herpesviruses and poxviruses, several of which have now been determined.

This nucleic acid-centered technology, in addition to its 'ultimate' achievement of nucleotide sequencing and the artificial manipulation of virus genomes, also offered significant advances in detection of viruses and virus infections involving nucleic acid hybridization techniques. There are many variants of this basic idea, but essentially, a hybridization probe, labeled in some fashion to facilitate detection, is allowed to react with a crude mixture of nucleic acids. The specific interaction of the probe sequence with complementary virus-encoded sequences, to which it binds by hydrogen-bond formation between the complementary base pairs, reveals the presence of the viral genetic material (Figure 1.7). In the past few years, this approach has been taken a stage further by the development of various *in vitro* nucleic acid amplification procedures, such as the polymerase chain reaction (PCR).

Liquid-phase hybribization

(a) Complex mixture of nucleic acids in solution

(b) Add labeled "probe" and allow hybridization with target sequence

(c) Remove unbound probe and analyze amount of label bound

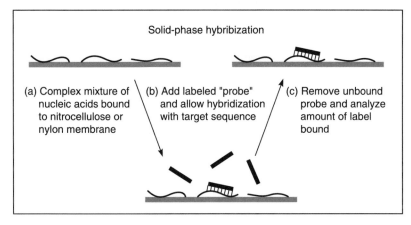

Solid-phase hybribization

(a) Complex mixture of nucleic acids bound to nitrocellulose or nylon membrane

(b) Add labeled "probe" and allow hybridization with target sequence

(c) Remove unbound probe and analyze amount of label bound

Figure 1.7 Nucleic acid hybridization relies on the specificity of base-pairing which allows a labeled nucleic acid probe to pick out a complementary target sequence from a complex mixture of sequences in the test sample. The label used to identify the probe may be a radioisotope or a non-isotopic label such as an enzyme or chemiluminescent system. Hybridization may be performed with both the probe and test sequences in the liquid phase (top figure) or with the test sequences bound to a solid phase, usually a nitrocellulose or nylon membrane. Both methods may be used to quantify the amount of the test sequence present, but solid-phase hybridization is also used to locate the position of sequences immobilized on the membrane. Plaque and colony hybridization are used to locate recombinant molecules directly from a mixture of bacterial colonies or bacteriophage plaques on an agar plate. 'Northern' and 'Southern' blotting are used to detect RNA and DNA respectively after transfer of these molecules from gels following separation by electrophoresis (cf. 'Western' blotting – Figure 1.2).

This is an even more sensitive technique, capable of detecting just a single molecule of viral nucleic acid (Figure 1.8).

More recently, there has also been renewed interest in virus proteins, based on a new biology itself dependent on manipulation of nucleic acids *in vitro* and advances in protein detection arising from immunology. Methods for *in vitro*

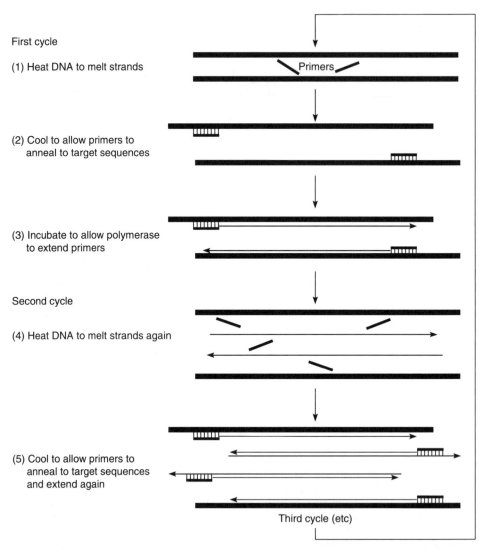

First cycle

(1) Heat DNA to melt strands

Primers

(2) Cool to allow primers to
anneal to target sequences

(3) Incubate to allow polymerase
to extend primers

Second cycle

(4) Heat DNA to melt strands again

(5) Cool to allow primers to
anneal to target sequences
and extend again

Third cycle (etc)

Figure 1.8 The polymerase chain reaction relies on the specificity of base-pairing between short synthetic oligonucleotide probes and complementary sequences in a complex mixture of nucleic acids to prime DNA synthesis using a thermostable DNA polymerase. Multiple cycles of primer annealing, extension, and thermal denaturation are carried out in an automated process, resulting in a massive amplification (2^n-fold increase after n cycles of amplification) of the target sequence located between the two primers.

synthesis and expression of proteins from molecularly cloned DNA have advanced rapidly, and many new analytical techniques are now available. Studies of protein–nucleic acid interactions are proving to be particularly valuable in understanding virus structure and gene expression.

In addition to examining the actual proteins of viruses, molecular biologists have one further trick up their sleeves. Because of the repetitive, essentially digitized nature of nucleotide sequences, computers have been found to be the ideal means of storing and processing this information. Moreover, they are capable of predicting, with reasonable accuracy, protein sequences (and detecting other functional regions) from nucleotide sequences (Figure 1.9). Although knowing the primary amino acid sequence of a polypeptide is a long way from understanding the subtleties of the protein molecule itself, this is still a powerful tool. Vast international databases of nucleotide and protein sequence information have now been compiled, and these can rapidly be consulted by computers to compare newly determined sequences with those whose function may have been studied in great detail.

Thus we have, in a sense, come full circle in our investigations of viruses – from particles via genomes back to proteins again, and have emerged with a far more

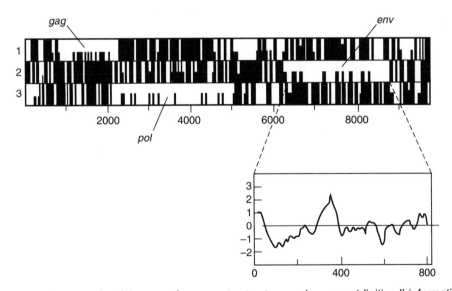

Figure 1.9 An example of the use of a computer to store and process 'digitized' information from a nucleic acid sequence. This figure shows an analysis of all of the open reading frames (ORFs) present in the (+)sense strand of an HIV-1 provirus. In each of the three frames, a line drawn completely across the frame represent a translation termination codon and a half line a methionine codon – a potential initiation site for translation of a protein. The ORFs present in the three main retrovirus genes, *gag*, *pol*, and *env*, can be seen. The computer has also been used to predict the amino acid sequence of the protein encoded by the *env* ORF and to analyze the hydrophobicity of the envelope protein (bottom of figure). This complex analysis took only a few minutes to perform using an ordinary personal computer. By hand, the same task would have taken several days.

profound understanding of these organisms. However, the present pace of research in virology tells us that there is still far more that we need to know.

Further Reading

Old, R.W. and Primrose, S.B. (1989). *Principles of Gene Manipulation* (4th edition). Blackwell Scientific, Oxford.

Roitt, I. (1988). *Essential Immunology* (6th edition). Blackwell Scientific, Oxford.

Sambrook, J., Fritsch, E.F. and Maniatis, T. (1989). *Molecular Cloning* (2nd edition). Cold Spring Harbor Press, Cold Spring Harbor.

Watt, I.M. (1985). *The Principles and Practice of Electron Microscopy*. Cambridge University Press, Cambridge.

Bibliography

Jaenisch, R. (1988). Transgenic animals. *Science* **240**: 1468–74.

Levine, A.J. (1992). *Viruses*. W.H. Freeman/Scientific American Library, New York.

Stanley, W.M. (1935). Isolation of a crystalline protein possessing the properties of tobacco mosaic virus. *Science* **81**: 644.

Waterson, A.P. and Wilkinson, L. (1978). *An Introduction to the History of Virology*. Cambridge University Press, Cambridge.

Chapter 2

Particles

The Function and Formation of Virus Particles

Why bother to form a virus particle to contain the genome? Indeed, some infectious agents, such as viroids, do not (see Chapter 8). However, the fact that viruses struggle with the genetic and biochemical burden entailed in encoding and assembling the components of a particle indicates that there must be positive benefits to this strategy. At the simplest level, the function of the outer shells of a virus particle is to protect the fragile nucleic acid genome from physical, chemical, or enzymatic damage. After leaving the host cell, the virus enters a hostile environment which would quickly inactivate the unprotected genome. Nucleic acids are susceptible to physical damage such as shearing by mechanical forces and to chemical modification by ultraviolet light (from sunlight). The natural environment is heavily laden with nucleases either derived from dead or leaky cells or deliberately secreted by vertebrates as defense against infection and in most cases, the breaking of a single phosphodiester bond in the backbone of a virus with a single-stranded genome, or chemical modification of one nucleotide is sufficient to inactivate that virus particle by rendering replication of the genome impossible. How is protection against this achieved? The protein subunits in a virus **capsid** are multiply redundant, i.e. present in many copies per particle. Damage to one or more subunits may render that particular subunit non-functional, but rarely does limited damage destroy the infectivity of the whole particle. This makes the capsid an effective barrier.

The outer surface of the virus is also responsible for recognition of and the first interaction with the host cell. Initially, this takes the form of binding of a specific virus-attachment protein to a cellular receptor molecule. However, the capsid also has a role to play in initiating infection by delivering the genome in a form in which it can interact with the host cell. In some cases this is a simple process which consists only of dumping the genome into the cytoplasm of the cell. In other cases this stage is much more complex; for example, retroviruses carry out extensive modifications to the virus genome while it is still inside the particle, converting two molecules of single-stranded RNA to one molecule of double-stranded DNA before delivering it to the cell nucleus. Hence, the role of the capsid is vital in allowing viruses to establish an infection.

To form an infectious particle, a virus must overcome two fundamental problems. First, it must assemble the particle utilizing the information available from the components which make up the particle themselves. Second, virus particles form regular geometric shapes, even though the proteins from which they are made are irregularly shaped. How do these simple organisms solve these difficulties? The solution to the first problem lies in the rules of symmetry. And so does the answer to the second.

Capsid Symmetry and Virus Architecture

It is possible to imagine a virus particle, the outer shell of which (the **capsid**) consists of a single, hollow protein molecule, which as it folds to assume its mature conformation, traps the virus genome inside. In practice, this arrangement cannot occur, for the following reason. The triplet nature of the genetic code means that three nucleotides (or base pairs in the case of viruses with double-stranded genomes) are necessary to encode one amino acid. Viruses cannot, of course, utilize an alternative, more economical, genetic code since this could not be deciphered by the host cell. Since the approximate molecular weight of a nucleotide triplet is 1000 and the average molecular weight of a single amino acid is 150, a nucleic acid can only encode a protein at most 15% of its own weight. Therefore, virus capsids must be made up of multiple protein molecules – subunit construction – and viruses must overcome the problem of how these subunits are to be arranged.

In 1955, Fraenkel-Conrat and Williams showed that when mixtures of purified tobacco mosaic virus (TMV) RNA and coat protein were incubated together, virus particles formed. The discovery that virus particles could form spontaneously from purified subunits without any extraneous information indicated that the particle was in the free energy minimum state and was therefore the favoured structure of the components. This stability is an important feature of the virus particle. Although some viruses are very fragile and unable to survive outside the protected host cell environment, many are able to persist for long periods, in some cases for years.

The forces which drive the assembly of virus particles include hydrophobic and electrostatic interactions – only rarely are covalent bonds involved in holding together the multiple subunits. In biological terms, this means that protein–protein, protein–nucleic acid, and protein–lipid interactions are used. It would be fair to say that the subtlety of these interactions is not fully understood for the majority of virus structures, but we now have a good understanding of general principles and repeated structural motifs which appear to govern the construction of diverse, unrelated viruses. These are discussed below under the two main classes of virus structures: helical and icosahedral symmetry.

Helical capsids

Tobacco mosaic virus (TMV) is representative of one of the two major structural classes seen in viruses, those with helical symmetry. The simplest way to arrange multiple, identical protein subunits is to use rotational symmetry and to arrange the irregularly shaped proteins around the circumference of a circle to form a disk. Multiple disks can then be stacked on top of one another to form a cylinder, with the virus genome coated by the protein shell or contained in the hollow center of the cylinder. Denaturation and phase-transition studies of TMV suggest that this is the form the particle takes (see Chapter 1).

Closer examination of the TMV particle by X-ray crystallography reveals that the structure of the capsid actually consists of a **helix** rather than a pile of stacked disks. A helix can be defined mathematically by two parameters: its amplitude (diameter) and pitch (the distance covered by each complete turn of the helix (Figure 2.1). Helices are rather simple structures formed by stacking repeated components with a constant relationship (amplitude and pitch) to one another. Note that if this simple constraint is broken, a spiral forms rather than a helix and this is quite unsuitable for containing a virus genome. In terms of each individual protein subunit, helices are described by the number of subunits per turn of the helix, μ, and the axial rise per subunit, p. The pitch of the helix, P, is therefore equal to:

$$P = \mu \times p$$

For TMV, $\mu = 16.3$, that is, there are 16.3 coat protein molecules per helix turn, and $p = 0.14$ nm. Therefore, the pitch of the TMV helix is $16.3 \times 0.14 = 2.28$ nm. TMV particles are rigid, rod-like structures, but some helical viruses demonstrate considerable flexibility and longer helical virus particles are often seen to be curved or bent. Flexibility is probably an important attribute. Long helical particles are likely to be subject to shear forces and the ability to bend reduces the likelihood of breakage or damage.

The fact that helical symmetry is a useful way of arranging a single protein

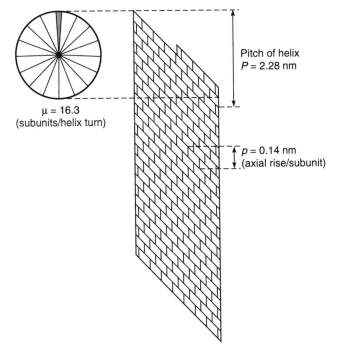

$\mu = 16.3$
(subunits/helix turn)

Pitch of helix
$P = 2.28$ nm

$p = 0.14$ nm
(axial rise/subunit)

Figure 2.1 Tobacco mosaic virus has a capsid consisting of many molecules of a single coat protein arranged in a constant relationship, forming a helix with a pitch of 22.8 Å.

subunit to form a particle is confirmed by the large number of different types of virus which have evolved with this capsid arrangement. Among the simplest helical capsids are those of the well-known bacteriophages of the family *Inoviridae*, such as M13 and fd – known as Ff phages. These phages are about 900 nm long and 9 nm in diameter and the particles contain five proteins (Figure 2.2). The major coat protein is the product of phage gene 8 (g8p) and there are 2700–3000 copies of this protein per particle, together with approximately five copies each of four minor capsid proteins, g3p, g6p, g7p and g9p, which are located at the ends of the filamentous particle. The primary structure of the major coat protein g8p explains many of the properties of the particle. Mature molecules of g8p consist of approximately 50 amino acid residues (a signal sequence of 23 amino acids is cleaved from the precursor protein during its translocation into the outer membrane of the host bacterium), and is almost entirely α-helical in structure so that the molecule forms a short rod. There are three distinct domains within this rod. There is a negatively charged region at the amino-terminal end which contains acidic

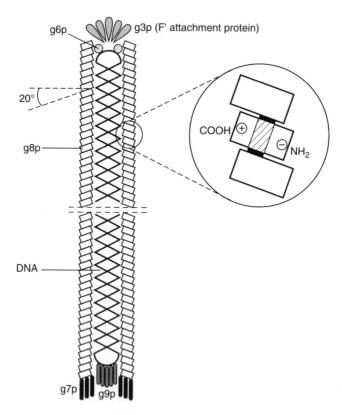

Figure 2.2 Schematic representation of the bacteriophage M13 particle. Major coat protein g8p is arranged helically, the subunits overlapping 'like the scales of a fish.' Other capsid proteins required for the biological activity of the virion are located at either end of the particle. *Inset* shows the hydrophobic interactions between the g8p monomers (shaded region).

amino acid residues and which forms the outer, hydrophilic surface of the virus particle, and a basic, positively charged region at the carboxy-terminal end which lines the inside of the protein cylinder adjacent to the negatively charged DNA genome. Between these two, there is a hydrophobic region which is responsible for interactions between the g8p subunits that allow the formation of and stabilize the phage particle (Figure 2.2). Ff phage particles are held together by the hydrophobic interactions between the coat protein subunits and this is demonstrated by the fact that the particles fall apart in the presence of chloroform, even though they do not contain any lipid component. The g8p subunits in successive turns of the helix interlock with the subunits in the turn below, are tilted at an angle of approximately 20° to the long axis of the particle and have been described as overlapping one another 'like the scales of a fish.' The value of μ (protein subunits per complete helix turn) is 4.5 and p (axial rise per subunit) = 1.5 nm.

Since the phage DNA is packaged inside the core of the helical particle, the length of the particle is dependent on the length of the genome. In all Ff phage preparations, 'polyphage' (containing more than one genome length of DNA), 'miniphage' (deleted forms containing 0.2–0.5 phage genomes length of DNA), and 'maxiphage' (genetically defective forms but containing more than one phage genome length of DNA) occur. This plastic property of these filamentous particles has been exploited by molecular biologists to develop the M13 genome as a cloning vector. Insertion of foreign DNA into the genome results in recombinant phage particles which are longer than the wild-type filaments. Unlike most viruses, there is no sharp cut-off genome length at which the genome can no longer be packaged into the particle. However, as M13 genome size increases, the efficiency of replication declines. While recombinant phage genomes 1–10% longer than the wild-type do not appear to be significantly disadvantaged, those 10–50% longer than the wild-type replicate significantly more slowly. Above 50% increase over the normal genome length it becomes progressively more difficult to isolate recombinant phage.

The structure of the Ff capsid also explains the events which occur on infection of suitable bacterial host cells. Ff phages are 'male-specific,' i.e. they require the F pilus on the surface of *Escherichia coli* for infection. The first event in infection is an interaction between g3p located at one end of the filament together with g6p, and the end of the F pilus. This interaction causes a conformational change in g8p. Initially, its structure changes from 100% α-helix to 85% α-helix and this causes the filament to shorten. The end of the particle attached to the F pilus flares open, exposing the phage DNA. Subsequently, a second conformational change in the g8p subunits reduces its α-helical content from 85% to 50%, causing the phage particle to form a hollow spheroid about 40 nm in diameter and expelling the phage DNA, thus initiating the infection of the host cell.

Many plant viruses show helical symmetry. These include members of the *Tobravirus, Tobamovirus, Furovirus, Hordeivirus, Potexvirus, Carlavirus, Capillovirus, Potyvirus* and *Closterovirus* groups. These particles vary from approximately 100 nm (*Tobravirus*) to approximately 1000 nm (*Closterovirus*) in length. The best studied example is, as stated above, TMV from the family *Tobamovirus*. Quite why so many groups of plant virus have evolved this structure is not clear, but it may be related either to the biology of the host plant cell, or alternatively, the way in which they are transmitted between hosts.

Helical, naked (i.e. non-enveloped) animal viruses do not exist. Once again, this probably reflects aspects of host cell biology and virus transmission, but the full reasons are not clear. There are a large number of animal viruses based on helical symmetry, but all have the addition of an outer lipid **envelope** (see below). There are too many viruses with this structure to list individually, but this category includes many of the best known human pathogens, such as influenzavirus (*Orthomyxoviridae*), mumps and measles viruses (*Paramyxoviridae*), and rabies virus (*Rhabdoviridae*). All possess single-stranded, negative-sense RNA genomes (see Chapter 3). The molecular design of all of these viruses is similar. The viral nucleic acid and a basic, nucleic acid-binding protein condense together in the infected cell to form a helical **nucleocapsid**. This protein:RNA complex serves to protect the fragile virus genome from physical and chemical damage and in some instances also provides other functions associated with virus replication. The envelope and its associated proteins are derived from the membranes of the host cell and are added to the nucleocapsid core of the virus during replication (see Chapter 4).

Some of these helical, enveloped animal viruses are relatively simple in structure, for example, rabies virus and the closely related virus, vesicular stomatitis virus (VSV) (Figure 2.3). The virus is built up around the negative-sense RNA genome, which in rhabdoviruses is about 11 000 nucleotides (11 kilobases, 11 kb) long. The

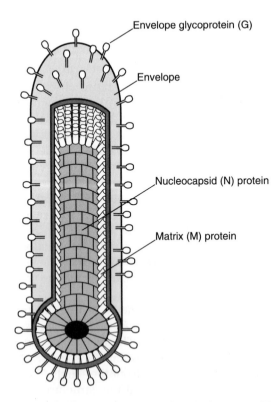

Envelope glycoprotein (G)

Envelope

Nucleocapsid (N) protein

Matrix (M) protein

Figure 2.3 Rhabdovirus particles, such as those of vesicular stomatitis virus, have an inner helical nucleocapsid surrounded by an outer lipid envelope and its associated glycoproteins.

RNA genome and basic nucleocapsid (N) protein interact to form a helical structure with a pitch of approximately 5 nm, which, together with two non-structural proteins, L and NS (which form the virus polymerase – see Chapter 4), makes up the core of the virus particle. There are 30–35 turns of the nucleoprotein helix in the core, which is about 180 nm long and 80 nm in diameter. The individual N protein monomers are approximately $9 \times 5 \times 3$ nm and each covers about nine nucleotides of the RNA genome. As in the case of the filamentous phage particles described above, the role of the N protein is to stabilize the RNA genome and protect it from chemical, physical and enzymatic damage. In common with most enveloped viruses, the nucleocapsid is surrounded by an amorphous layer, which interacts with both the core and the overlying lipid envelope linking them together. This is known as the matrix. The matrix (M) protein is usually the most abundant protein in the virus particle; for example, there are approximately 1800 copies of the M protein and approximately 1250 copies of the N protein in VSV particles. The lipid envelope and its associated proteins are discussed in more detail below.

It is clear that many different groups of viruses have evolved around helical symmetry. Simple viruses with small genomes use this architecture to provide protection for the genome without the need to encode multiple capsid proteins. More complex viruses utilize this structure as the basis of the virus particle, but elaborate on it with additional layers of protein and lipid.

Icosahedral (isometric) capsids

An alternative way of building a virus capsid is to arrange protein subunits in the form of a hollow quasispherical structure, enclosing the genome within. The criteria for arranging subunits on the surface of a solid are a little more complex than those for building a helix. In theory, a number of solid shapes can be constructed from repeated subunits, e.g. a tetrahedron (4 triangular faces), cube (6 square faces), octahedron (8 triangular faces), dodecahedron (12 pentagonal faces) and an **icosahedron**, a solid shape consisting of 20 triangular faces arranged around the surface of a sphere (Figure 2.4).

Early in the 1960s, direct examination of a number of small 'spherical' viruses by electron microscopy revealed that they appeared to have icosahedral symmetry. At first sight, it is not obvious why this pattern should have been 'chosen' by diverse virus groups. However, although in theory it is possible to construct virus capsids based on simpler symmetrical arrangements, such as tetrahedra or cubes, there are practical reasons why this does not occur. As described above, it is more economic in terms of genetic capacity to design a capsid based on a large number of identical, repeated protein subunits rather than fewer, larger subunits. It is unlikely that a simple tetrahedron consisting of four identical protein molecules would be large enough to contain even the smallest virus genome. If it were, it is probable that the gaps between the subunits would be so large that the particle would be leaky, and fail to carry out its primary function of protecting the virus genome.

In order to construct a capsid from repeated subunits, a virus must 'know the rules' which dictate how these are arranged. For an icosahedron the rules are

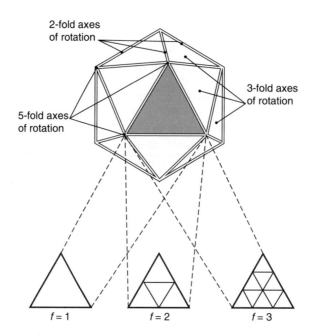

Figure 2.4 Illustration of the 2–3–5 symmetry of an icosahedron. More complex ('higher order') icosahedra can be defined by the triangulation number of the structure, $T = f^2 \times P$. Regular icosahedra have faces consisting of equilateral triangles and are formed when the value of $P = 1$ or 3. All other values of P give rise to more complex structures with either a left-hand or right-hand skew.

based on the rotational symmetry of the solid, known as 2–3–5 symmetry, which has the following features (Figure 2.4):

- an axis of two-fold rotational symmetry through the center of each edge;
- an axis of three-fold rotational symmetry through the center of each face; and
- an axis of five-fold rotational symmetry through the center of each corner (vertex).

Since protein molecules are irregularly shaped and are not regular equilateral triangles, the simplest icosahedral capsids are built up by using three identical subunits to form each triangular face. This means that 60 identical subunits are required to form a complete capsid. A few simple virus particles are constructed in this way; for example, bacteriophages of the family *Microviridae*, such as ϕX174, the complete structure of which has recently been determined by X-ray crystallography.

In most cases, analysis reveals that icosahedral virus capsids contain more than 60 subunits, for the reasons of genetic economy given above. This presents a difficulty. A regular icosahedron composed of 60 identical subunits is a very stable structure, since all the subunits are equivalently bonded, i.e. they show the same spacing relative to one another and each occupies the minimum free energy state. With more than 60 subunits it is impossible for them all to be arranged completely symmetrically with exactly equivalent bonds to all their neighbors, since a true regular icosahedron consists of only 20 subunits. To solve this problem, in 1962 Caspar and Klug proposed the idea of **quasi-equivalence**. Their simple idea was that subunits in *nearly* the same local environment form *nearly* equivalent bonds with their neighbors, permitting self-assembly of icosahedral capsids from multiple subunits. In the case of these higher order icosahedra, the symmetry of the particle is defined by the **triangulation number** of the icosahedron (Figure 2.4). The triangulation number, T, is defined by:

$$T = f^2 \times P$$

where f is the number of subdivisions of each side of the triangular face; f^2 is therefore the number of subtriangles on each face; and $P = h^2 + hk + k^2$, where h and k are any distinct, non-negative integers. This means that values of T fall into the series 1, 3, 4, 7, 9, 12, 13, 16, 19, 21, 25, 27, 28, etc. When $P = 1$ or 3, a regular icosahedron is formed. All other values of P give rise to icosahedra of the 'skew' class – the subtriangles making up the icosahedron are not symmetrically arranged with respect to the edge of each face. Detailed structures of icosahedral virus particles with $T = 1$ (*Microviridae*, e.g. ϕX174), $T = 3$ (many insect, plant, and animal viruses – see below), $T = 4$ (togaviruses) and $T = 7$ (the heads of the tailed bacteriophages such as λ) have all been determined. Virus particles with still larger triangulation numbers use different kinds of subunit assemblies for the faces and vertices of the icosahedron and have internal scaffolding proteins which act as a framework. These direct the assembly of the capsid, typically by bringing together pre-formed subassemblies of proteins.

The capsids of picornaviruses (*Picornaviridae*) provide a good illustration of the construction of icosahedral virus particles. In recent years, the atomic structure of the capsids of a number of different picornaviruses have been determined. These include poliovirus types 1 and 3 (PV1 and PV3), foot-and-mouth disease virus (FMDV) and human rhinovirus type 14 (HRV-14) as well as a number of others. In fact, the structure of these virus particles is remarkably similar to those of many other unrelated viruses, such as insect viruses of the family *Nodaviridae* and plant viruses from the *Comovirus* group. All these virus groups have icosahedral capsids approximately 30 nm in diameter with triangulation number $T = 3$ (Figure 2.5). The capsid is composed of 60 repeated subassemblies of proteins, each containing three major subunits, VP1, VP2, and VP3. This means that there are $60 \times 3 = 180$ surface monomers in the whole picornavirus particle. All three proteins are based on a similar structure, consisting of 150–200 amino acid residues in what has been described as an '8-strand anti-parallel β-barrel.'

Knowledge of the structure of these picornavirus capsids also reveals information about the way in which they are assembled and the function of the mature capsid. In fact, picornavirus capsids contain four structural proteins. In addition to the three major proteins VP1–3 (above), there is a small fourth protein, VP4. VP4 is located predominantly on the inside of the capsid and is not exposed at the surface of the particle. The way in which the four capsid proteins are processed from the

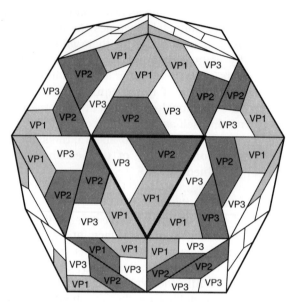

Figure 2.5 Picornavirus particles are icosahedral structures with a triangulation number *T* = 3. Three virus proteins, VP1, 2, and 3, comprise the surface of the particle. A fourth protein, VP4 is not exposed on the surface of the virion but is present in each of the 60 repeated units which make up the capsid.

initial **polyprotein** (see Chapter 5) has long been known from biochemical studies of picornavirus-infected cells. VP4 is formed from cleavage of the VP0 precursor into VP2+VP4 late in assembly and is myristylated at its amino terminus, i.e. it is modified after translation by the covalent attachment of myristic acid, a 14-carbon unsaturated fatty acid. Five VP4 monomers form a hydrophobic micelle, driving the assembly of a pentameric subassembly. There is biochemical evidence that these pentamers, which form the vertices of the mature capsid, are a major precursor in the assembly of the particle. Hence, the chemistry, structure, and symmetry of the proteins which make up the picornavirus capsid reveal how assembly is driven.

Because they are the cause of a number of important human diseases, picornaviruses have been intensively studied by virologists over the last decade or so. This interest has resulted in an outpouring of knowledge about these structurally simple viruses. Detailed knowledge of the structure and surface geometry of HRV-14 has revealed much about the interaction of rhinoviruses with their host cells and with the immune system. In recent years, much has been learned not only about this virus, but also about the identity of its cellular receptor, ICAM-1 (see below and Chapter 4). In addition, the immunological structure of the picornavirus particle has also been elucidated. A number of studies have been published using panels of monoclonal antibodies whose binding sites have been 'mapped' to the primary amino acid sequence of the virus by examining their reactivity towards mutant viruses or by the use of synthetic peptides to block binding. The information from these experiments has been used to identify a number of discrete antibody-neutralization sites on the surface of the virus particle. Some of these correspond to contiguous linear regions of the primary amino acid sequence of the capsid proteins, others, known as conformational sites, result from the bringing together of separated stretches of amino acids in the mature virus. With the elucidation of detailed picornavirus capsid structures, these regions have now been physically identified on the surface of the particle. They correspond mostly to hydrophilic, exposed loops of amino acid sequence, readily accessible to antibody binding, which are repeated on each of the pentameric subassemblies of the capsid. Now that the physical constraints on these sites are known, this type of information is being used to artificially manipulate them, even to build 'antigenic chimeras' with the structural properties of one virus but expressing crucial antigenic sites from another. With the application of the rational, computer-aided design tools now available, it is probable that the efficiency with which this type of chimera can be designed and built should increase. Indeed, these compound viruses may turn out to be the vaccines of the future.

Enveloped Viruses

So far, this chapter has concentrated on the structure of 'naked' virus particles, i.e. those in which the capsid proteins are exposed to the external environment. Such viruses are produced from infected cells at the end of the replicative cycle when the cell dies, breaks down and lyzes, releasing the virions which have been built up internally. This simple strategy has drawbacks. In some circumstances it is wasteful, resulting in the premature death of the cell. It also reduces the possibilities for persistent or latent infections. Therefore, many viruses have devised strategies to effect an exit from the infected cell without its total destruction. The difficulty this presents is that all living cells are covered by a membrane composed of a lipid bilayer. The viability of the cell depends on the integrity of this membrane. Viruses leaving the cell must, therefore, allow this membrane to remain intact. This is achieved by extrusion (**budding**) of the particle through the membrane, during which process the particle becomes coated in a lipid **envelope** derived from the host cell membrane and with a similar composition (Figure 2.6).

Viruses have also turned this necessity into an advantage. The structure underlying the envelope may be based on helical or icosahedral symmetry and may be formed before or as the virus leaves the cell. In the majority of cases, enveloped viruses use cellular membranes as sites allowing them to direct assembly. The formation of the particle inside the cell, maturation and release are in many cases a continuous process. The site of assembly varies for different viruses. Not all use the cell surface membrane; many use cytoplasmic membranes such as the Golgi apparatus, others, such as herpesviruses, which replicate in the nucleus may utilize the nuclear membrane. In these cases, the virus is usually extruded into some form of vacuole, in which it is transported to the cell surface and subsequently released. These points are discussed in more detail in Chapter 4.

If the virus particle became covered in a smooth, unbroken lipid bilayer, this would be its undoing. Such a coating is effectively inert, and though effective as a protective layer preventing desiccation of or enzymatic damage to the particle, it

Figure 2.6 Enveloped virus particles are formed by budding through a host cell membrane, during which the particle becomes coated with a lipid bilayer derived from the cell membrane. For some viruses, assembly of the structure of the particle and budding occur simultaneously, whereas in others a pre-formed core pushes out through the membrane.

would not permit recognition of receptor molecules on the host cell. Therefore, viruses modify their lipid envelopes by the synthesis of several classes of proteins which are associated in one of three ways with the envelope (Figure 2.7). These can be summarized as follows:

- *Matrix proteins*: These are internal virion proteins whose function is effectively to link the internal nucleocapsid assembly to the envelope. Such proteins are

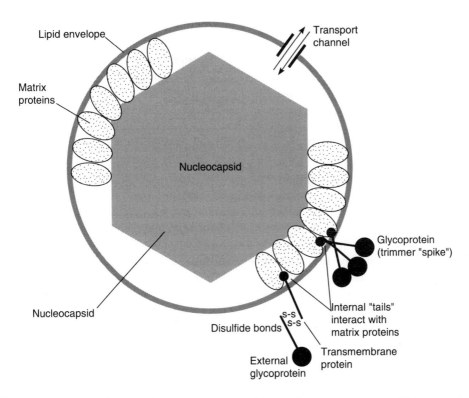

Figure 2.7 Several classes of proteins are associated with virus envelopes. Matrix proteins link the envelope to the core of the particle. Virus-encoded glycoproteins inserted into the envelope serve several functions. External glycoproteins are responsible for receptor recognition and binding, while transmembrane proteins act as transport channels across the envelope. Host-cell derived proteins are also sometimes found to be associated with the envelope, usually in small amounts.

not usually glycosylated and are often very abundant, for example, in retroviruses they comprise approximately 30% of the total weight of the virion. Some matrix proteins contain transmembrane anchor domains, others are associated with the membrane by hydrophobic patches on their surface or by protein:protein interactions with envelope glycoproteins.

- *Glycoproteins*: These are transmembrane proteins anchored to the membrane by a hydrophobic domain, and can be subdivided into two types by their function:
  - (a) *External glycoproteins* are anchored in the envelope by a single transmembrane domain. Most of the structure of the protein is on the outside of the membrane, with a relatively short internal tail. Often, individual monomers associate to form the 'spikes' visible in electron micrographs on the surface of many enveloped viruses. Such proteins are the major antigens of enveloped viruses. The glycosylation is either *N*- or *O*-linked, and many of these proteins are heavily glycosylated; up to 75% of the protein by weight may consist of

post-transcriptionally added sugar groups. These proteins are usually the major antigens of enveloped viruses and provide contact with the external environment, frequently serving a number of important functions; for example, influenza virus hemagglutinin is required for receptor binding, membrane fusion, and hemagglutination.

(b) *Transport channel proteins* contain multiple hydrophobic transmembrane domains, forming a protein-lined channel through the envelope. This enables the virus to alter the permeability of the membrane (e.g. ion channels). Such proteins are often important in modifying the internal environment of the virion, permitting or even driving biochemical changes necessary for maturation of the particle and the development of infectivity (e.g. influenza-virus M_2 protein).

While there are many enveloped vertebrate viruses, there are only a few enveloped plant viruses, and most of these belong to the *Rhabdoviridae* family, whose structure has already been discussed (see Helical Symmetry). Aside from plant rhabdoviruses, only a few bunyaviruses which infect plants and the members of the *Tospovirus* group have outer lipid envelopes. This relative paucity of enveloped plant viruses probably reflects aspects of host cell biology, in particular the mechanism of release of the virus from the infected cell, which requires a breach in the rigid cell wall. This constraint does not apply to viruses of prokaryotic organisms, where there are a number of enveloped virus families (*Plasmaviridae*, the SSV-1 group, *Lipothrixviridae*, and *Cystoviridae*).

Complex Viruses

The majority of viruses can be fitted into one of the three structural classes outlined above, i.e. those with helical symmetry, icosahedral symmetry, or enveloped viruses. However, there are many viruses whose structure is more complex. In these cases, although the general principles of symmetry already described are often used to build part of the virus shell (this term being appropriate here since such viruses often consist of several layers of protein and lipid), the larger and more complex viruses cannot be simply defined by a mathematical equation as can a simple helix or icosahedron. Because of the complexity of some of these viruses, they have defied attempts to determine detailed atomic structures using the techniques described in Chapter 1.

An example of such a group is the *Poxviridae* family. These viruses have oval or 'brick-shaped' particles 200–400 nm long. In fact, these particles are so large that they were first observed in vaccine lymph using high-resolution optical microscopes in 1886, and thought at that time to be 'the spores of micrococci.' The external surface of the virion is ridged in parallel rows, sometimes arranged helically. The particles are extremely complex and have been shown to contain more than 100 different proteins. During replication, two forms of particle are observed: extracellular forms which contain two membranes and intracellular particles which only have an inner membrane (Figure 2.8). Thin sections under the electron microscope reveal that the outer surface of the virion is composed of lipid

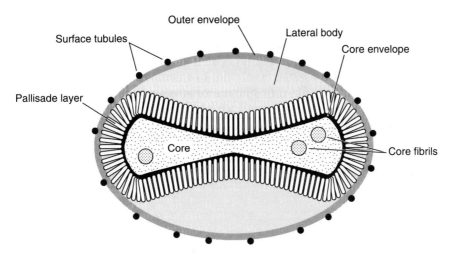

Outer envelope

Surface tubules

Lateral body

Core envelope

Pallisade layer

Core

Core fibrils

Figure 2.8 Poxvirus particles are the most complex virions known and contain more than a hundred virus-encoded proteins, arranged in a variety of internal and external structures.

and protein. This surrounds the core, which is biconcave (dumbbell-shaped), and two 'lateral bodies' whose function is unknown. The core is composed of a tightly compressed nucleoprotein and the double-stranded DNA genome is wound around it. Antigenically, poxviruses are very complex, inducing both specific and cross-reacting antibodies – hence the possibility of vaccinating against one disease with another virus (e.g. the use of vaccinia virus to immunize against smallpox (variola) virus). Poxviruses and a number of other complex viruses also emphasize the true complexity of some viruses – there are at least ten enzymes present in poxvirus particles, mostly concerned with nucleic acid metabolism/genome replication.

Poxviruses form the most complex particles known and even though the complete nucleotide sequence of the genome of at least two representatives of this family (smallpox and vaccinia virus) has now been determined (see Chapter 3), a complete elucidation of the structure of these particles has not yet been achieved. This is an extreme case in virology, included here as a counterbalance to the description of some of the simplest viruses given above. In other cases, the particle structure of complex viruses has been much more completely investigated. One of the prime examples of such a group are the tailed phages of *Escherichia coli*. These viruses, members of the families *Myoviridae*, *Siphoviridae* and *Podoviridae*, have been extensively studied for excellent reasons – they are easy to propagate in bacterial cells, they can be obtained in high titers, and they are easily purified, facilitating biochemical and structural studies. The head of the particles consists essentially of an icosahedral shell with $T = 7$ symmetry, attached via a collar to a contractile, helical tail. At the end of the tail is a plate which functions in attachment to the bacterial host and also in penetration of the bacterial cell wall by virtue of lysozyme-like enzymes associated with the plate. In addition to these structures, thin protein fibers are attached to the plate, which along with the tail plate, are involved in binding to the receptor molecules in the wall of the host cell. In detail, the structure of these phages is rather more complex than this simple picture; for example, there are several internal proteins and polyamines associated with the genomic DNA in the head and an internal tube structure inside the outer sheath of the helical tail. In the infected bacterial cell, there are separate assembly pathways for the head and tail sections of the particle, which come together at a late stage to make up the virion (Figure 2.9). Therefore, these viruses illustrate how complex particles can be built up from the simple principles outlined above. An even clearer example of this phenomenon is provided by the structure of geminivirus particles, which consist of two twinned $T = 1$ icosahedra. Each icosahedron has one morphological subunit missing and the icosahedra are joined at the point such that the mature particle contains 110 protein monomers arranged in 22 morphological subunits.

Members of the reovirus genus have non-enveloped, icosahedral (probably $T =$

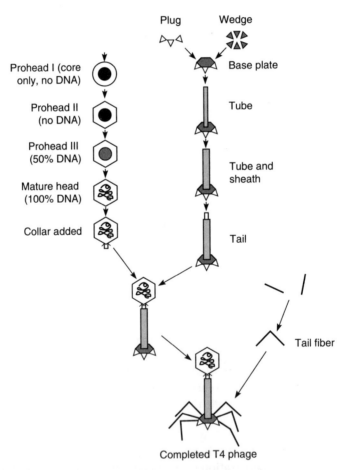

Figure 2.9 Simplified version of the assembly of bacteriophage T4 particles. The head and tail sections are assembled separately and are brought together at a relatively late stage. This complex process was painstakingly worked out by the isolation of phage mutants in each of the virus genes involved. In addition to the major structural proteins, a number of minor 'scaffolding' proteins are involved in guiding the formation of the complex particle.

13) capsids composed of double protein shell with a complex structure (Figure 2.10). The structure of orbivirus capsids, such as bluetongue virus, is slightly smaller in diameter but apparently made up following a similar $T = 13$ design. The outer shell is approximately 80 nm in diameter and the inner shell, or core, about 60 nm. The structure of the outer shell is complex and has not been completely determined. The double-stranded RNA genome is packed tightly inside the core. Notably, 12 'spikes' protrude out from the core through the outer shell. The overall composition of *Orthoreovirus* particles is as follows:

σ1, σ3, μ1c   – in the outer capsid
λ1, λ2, λ3, σ2, μ2 – in the virus core
μNS, σNS, σ1s   – non-structural proteins (enzymes which transcribe the
                      double-stranded RNA genome).

After many years of study and the isolation of a very large number of mutants, there is now probably more detail known about the structure:function relationships of the reovirus capsid than that of any other virus, with the possible exception of the picornaviruses. These particles are very stable in the environment – this is notably the case with the rotaviruses, which are commonly spread through contaminated drinking water. Rotavirus particles in fecal material stored at room

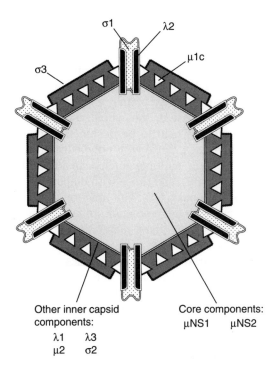

Figure 2.10 Reovirus particles consist of an icosahedral, double-shell arrangement of proteins surrounding the core.

temperature are able to retain their infectivity for 7 months and are not very susceptible to chlorine-containing disinfectants, thus demonstrating the efficiency with which these complex particles protect the fragile RNA genome.

Recently, the complex structure of cauliflower mosaic virus (CaMV) has been determined to 3 nm resolution by the use of newer techniques of structural investigation. Although this level of resolution is one order of magnitude less refined than that which can be achieved by X-ray diffraction, this report represents a considerable advance. The structure of CaMV has been a longstanding mystery since the virus is very fragile and it is distorted by drying, sectioning, or negative-staining. By examining electron-density maps of frozen-hydrated virus preparations and using sophisticated image reconstruction techniques, the CaMV virion was found to be 54 nm in diameter, and to consist of three concentric protein layers with an inner solvent-filled cavity of 27 nm diameter. The outer shell is comprised of 72 capsomers (12 pentamers and 60 hexamers) arranged with $T = 7$ symmetry, making a total of 420 individual subunits. Interestingly, the 8 kbp double-stranded DNA genome (see Chapter 3) is distributed in the inner layers II and III rather than in the internal cavity.

The final example of complex virus structure to be considered will be the baculovirus family (Figure 2.11). In recent years, these viruses have attracted much interest for a number of reasons. They are natural pathogens of arthropods and naturally occurring as well as genetically manipulated baculoviruses are under active investigation as biological control agents for insect pests. In addition, occluded baculoviruses (see below) are increasingly being used as expression vectors to produce large amounts of recombinant proteins. These complex viruses contain 12–30 structural proteins, and consist of a rod-like (hence '*baculo*') nucleocapsid of 30–60 nm diameter and 250–300 nm in length which contains the 88–160 kbp double-stranded DNA genome. The nucleocapsid is surrounded by an envelope, outside of which there may or may not be a crystalline protein matrix. If this outer protein shell is present, the whole assemblage is referred to as an 'occlusion body' and the virus is said to be occluded (Figure 2.11). There are two genera of occluded baculoviruses, those with polyhedral occlusions 1000–15 000 nm in diameter which may contain multiple nucleocapsids within the envelope (e.g. nuclear polyhedrosis virus – NPV), and those with ellipsoidal occlusions 200–500 nm in diameter (e.g. granulosis viruses – GV). The function of these enormous occlusion bodies is to confer resistance to adverse environmental conditions, which enables the virus to persist in soil or on plant materials for extended periods of time waiting to be ingested by a new host.

The effectiveness of this strategy is such that these viruses can be regarded as being literally armour-plated! Interestingly, the strategy of producing occluded particles appears to have evolved independently in at least three groups of insect viruses. In addition to the baculoviruses, occuluded particles are also produced by insect reoviruses (cytoplasmic polyhedrosis viruses) and poxviruses (entemopoxviruses). However, this resistant coating would be the undoing of the virus if it were not removed at an appropriate time to allow replication to proceed. This is achieved by virtue of the fact that the occlusion body is alkali-labile and dissolves in the high pH environment of the insect midgut, releasing the nucleocapsid and allowing it to infect the host. Although the structure of the whole particle has

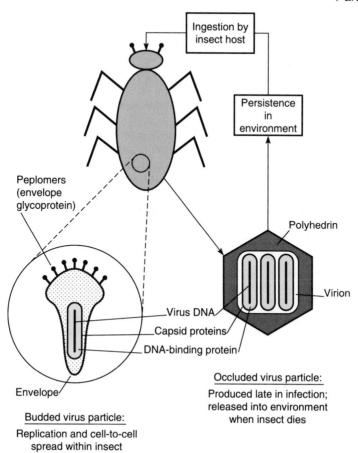

Figure 2.11 Some baculovirus particles exist in two forms: a relatively simple 'budded' form found within the host insect, and a crystalline, protein-occluded form responsible for environmental persistence.

not been completely determined, it is known that the occlusion body is composed of many copies of a single protein of approximately 245 amino acids, polyhedrin. To form the occlusion body, this single gene product is hyper-expressed late in infection by a very strong transcriptional promoter and it is by virtue of cloning genes to be expressed in the polyhedrin gene that baculoviruses have been manipulated as expression vectors.

Protein–Nucleic Acid Interactions and Genome Packaging

The primary function of the virus particle is to contain and protect the genome before delivering it to the appropriate host cell. Therefore, it is clear that the proteins of the capsid must interact with the nucleic acid genome. However, once again the physical constraints of incorporating a relatively large nucleic acid molecule into a relatively small capsid present considerable problems which must be overcome. In most cases, the linear virus genome when stretched out in solution is at least an order of magnitude longer than the diameter of the capsid. Merely folding the genome in order to stuff it into such a confined space is quite a feat of topology in itself, but is compounded by the fact that repulsion by the cumulative negative electrostatic charges on the phosphate groups of the nucleotide backbone mean that the genome resists being crammed into a small space. Viruses overcome this difficulty by packaging along with the genome a number of positively charged molecules in order to counteract this negative charge repulsion. These include small, positively charged ions (Na, Mg, K, etc.), polyamines and various nucleic acid-binding proteins. Some of these latter proteins are virus-encoded and contain amino acids with basic side-chains such as arginine and lysine which interact with the genome. There are many examples of such proteins, for example, retrovirus NC and rhabdovirus N (nucleocapsid) proteins, and influenzavirus NP protein (nucleoprotein). Many viruses with double-stranded DNA genomes have basic histone-like molecules closely associated with the DNA. Again, some of these are virus-encoded, e.g. adenovirus polypeptide VII. In other cases, however, the virus may utilize cellular proteins; for example, the polyomavirus genome assures a chromatin-like structure in association with four cellular histone proteins, H2A, H2B, H3 and H4, similar to that of the host cell genome.

The second problem the virus must overcome is how to achieve the specificity required to select and encapsidate the virus genome from the large background of cellular nucleic acids. In most cases, by the late stages of virus infection when assembly of virus particles occurs (see Chapter 4), transcription of cellular genes has been reduced and a large pool of virus genomes have accumulated. The overproduction of viral nucleic acids eases but does not eliminate the problem of specific genome packaging. Therefore, a specific virus-encoded capsid or nucleo-capsid protein is required to achieve this end and many viruses, even those with relatively short, compact genomes such as retroviruses and rhabdoviruses, encode this type of protein.

On the other side of the packaging equation are the specific nucleotide sequences in the genome (the **packaging signal**) which permit the virus to select genomic nucleic acids from the cellular background. The packaging signal from a number of virus genomes has been identified. Examples are the 'Ψ' signal in murine retrovirus genomes which has been used to package synthetic 'retrovirus vector' genomes into a virus particle, and the sequences responsible for packaging the genomes of several DNA virus genomes (some adenoviruses and herpesviruses) which have been clearly and unambiguously defined. However, it is clear from a number of different approaches that accurate and efficient genome packaging requires information not only from the linear nucleotide sequence of the genome, but also from regions of secondary structure formed by the folding of the genomic nucleic acid into complex forms. In many cases, attempts to find a unique, linear packaging signal in virus genomes have failed. The probable reason for this is that the key to the specificity of genome packaging in most viruses lies in the secondary structure of the genome.

Viruses with segmented genomes (see next chapter) face further problems: they must not only encapsidate only viral nucleic acid and exclude host cell molecules, but they must also attempt to package one of each of the required genome segments. It is important to realize that during assembly, viruses frequently make mistakes. These can be physically measured by particle:infectivity ratios, i.e. the ratio of the total number of particles in a virus preparation (counted by electron microscopy) to the number of particles able to give rise to infectious progeny (measured by plaque or limiting dilution assays). This value is in some cases found to be several thousand particles to each infectious virion and only rarely approaches a ratio of 1:1. However, a few simple calculations show that viruses such as influenza have far lower particle:infectivity ratios than could be achieved by random packaging of eight distinct genome segments. It is now believed that each influenza virus particle contains more than eight RNA segments. This redundancy is sufficient to ensure that a reasonable proportion of virus particles in the population will contain at least one of each of the eight segments and will thus be infectious.

Like many other aspects of virus assembly, the way in which packaging is controlled is not well understood but, as stated above, the key must lie in the specific molecular interactions between the genome and the capsid. In general, the physical structure of virus genomes within virus particles has been poorly studied, although the genetics of packaging have been extensively investigated in recent years. This is a pity, because it is unlikely that we will be able fully to appreciate this important aspect of virus replication without this information, but is understandable because the techniques used to determine the structure of virus capsids (e.g. X-ray diffraction) only rarely reveal any information about the state of the genome within its protein shell. However, there are at least two cases where something is known about the mechanism and specificity of genome encapsidation: one is a virus with helical symmetry (M13) and one is an icosahedral virus (bean pod mottle virus – BPMV – a member of the *Comovirus* group).

The protein–nucleic acid interactions in the M13 particle are relatively simple to understand (Figure 2.2). The primary sequence of the g8p molecule determines the orientation of the protein in the capsid. In simple terms, the inner surface of the rod-like phage capsid is positively charged and interacts with the negatively charged genome, while the outer surface of the cylindrical capsid is negatively charged. However, the way in which the capsid protein and genome are brought together is a little more complex than this. During replication, the genomic DNA is associated with a non-structural DNA-binding protein, g5p. This is the most abundant of all virus proteins in the infected *E. coli* cell, and coats the newly replicated single-stranded phage DNA, forming an intracellular rod-like structure similar to the mature phage particle, but somewhat longer and thicker (1100×16 nm). The function of this protein is to protect the genome from host cell nucleases and to interrupt genome replication, sequestering newly formed strands as substrates for encapsidation. Newly synthesized coat protein monomers (g8p) are associated with the inner (cytoplasmic) membrane of the cell and it is at this site that assembly of the virus particle occurs. The g5p coating is stripped off as the particle passes out through the membrane and is essentially exchanged for the mature g8p coat (plus the accessory proteins). The forces which drive this process are not fully understood, but the protein–nucleic acid interactions which occur appear to be rather simple and involve opposing electrostatic charges and the stacking of the DNA bases between the planar side-chains of the proteins. This is confirmed by the plasticity of the M13 genome and its ability to freely encapsidate extra genetic material.

Protein–nucleic acid interactions in other helical viruses, such as rhabdoviruses, are rather more complex. In most enveloped helical viruses, a nucleoprotein core forms first and this is then coated by matrix proteins and the envelope and its associated glycoproteins (Figure 2.3). The fine structure of the core has not been determined, but appears to show cross-striations 4.5–5.0 nm apart, each of these presumably equating to one turn of the protein–RNA complex (rather like TMV). The matrix protein shows an apparently hexagonal pattern and it is not clear how this is related to the structure of the underlying nucleocapsid, or how the rounded end of the virus particle is formed.

Even less is known about the arrangement of the genome inside virus particles with icosahedral symmetry. There are, however, now a few exceptions to this statement. The first is BPMV, a comovirus with a bipartite genome, the structure of which is based on a $T = 3$ icosahedron similar to that of picornaviruses (above). X-ray crystallography has shown that the RNA is folded in such a way that it assumes icosahedral symmetry, corresponding to that of the capsid surrounding it. The regions which contact the capsid proteins are single-stranded and appear to interact by electrostatic forces rather than covalent bonds. Recently, the atomic structure of φX174 has also shown that a portion of the DNA genome interacts with two arginine residues exposed on the inner surface of the capsid in a similar way to BPMV.

Therefore, a consensus of opinion about the physical state of nucleic acids within icosahedral virus capsids appears to be emerging. Just as the icosahedral capsids of many genetically unrelated viruses are based on monomers with a common '8-strand anti-parallel β-barrel' structural motif, the genomes inside also appear to display icosahedral symmetry, the vertices of which interact with basic amino acid residues on the inner surface of the capsid. Thus, these common structural motifs may in time explain how viruses selectively package the required genomic nucleic acids and may even offer opportunities to design specific drugs to inhibit these interactions.

LEEDS METROPOLITAN UNIVERSITY LIBRARY

Virus Receptors – Recognition and Binding

Cellular receptor molecules used by a number of different viruses from diverse taxonomic groups have now been identified. These include the human immunodeficiency virus receptor, CD4, and the Epstein–Barr virus receptor, CR2, both of which normally participate in cell–cell recognition and signaling. The interaction of a virus with a cellular receptor is a major event in determining the subsequent events in replication and the outcome of infections. It is this interaction which activates inert extracellular virus particles and initiates the replication cycle. Receptor binding is considered in detail in Chapter 4.

Other Interactions of the Virus Capsid with the Host Cell

As stated at the beginning of this chapter, the function of the virus capsid is not only to protect the genome, but also to deliver it to a suitable host cell (below) and, more specifically, the appropriate compartment of the host cell (in the case of eukaryote hosts) to allow replication to proceed. One example is the nucleocapsid proteins of viruses which replicate in the nucleus of the host cell. These molecules contain within their primary amino acid sequences 'nuclear localization signals' which are responsible for the migration of the virus genome plus its associated proteins into the nucleus where replication can occur. Again, these events are discussed in Chapter 4.

Virions are not inert structures. Many virus particles contain one or more enzymatic activities, although in most cases these are not active outside the biochemical environment of the host cell. All viruses with negative-sense RNA genomes must carry with them a virus-specific RNA-dependent RNA polymerase, since uninfected cells have no mechanism for RNA-dependent RNA polymerization and, therefore, genome replication could not occur if this enzyme were not included in the virus particle. The more complex DNA viruses (e.g. herpesviruses and poxviruses) carry a multiplicity of enzymes, mostly concerned with nucleic acid metabolism in some way. Reverse transcription of retrovirus genomes occurs inside a particulate complex and not free in solution.

Summary

This chapter is in no way intended to be a complete list of all the virus structures which are now known, but it attempts to illustrate with examples some of the principles that control the assembly of viruses and the difficulties of studying these minute structures. Readers must refer to scientific articles for the fine detail of known virus structures and for the many which will undoubtedly appear in the next few years. Nevertheless, it is clear that there are a number of repeated structural motifs found in many different virus groups. The most obvious is the division of many virus structures into those based on helical or icosahedral symmetry. More subtly, common protein structures such as the '8-strand anti-parallel β-barrel' structural motif found in many $T = 3$ icosahedral virus capsids and the icosahedral folded RNA genome present inside some of these viruses are beginning to emerge. Virus particles are not inert. Many are armed with a variety of enzymes which carry out a range of complex reactions, most frequently concerned with the replication of the genome.

Further Reading

Harrison, S.C. (1990). Principles of virus structure. In *Field's Virology* (2nd edition), Eds B.N. Fields and D.M. Knipe. Raven Press, New York, chap. 3, pp. 37–61.

Klug, A. (1983). Architectural design of spherical viruses. *Nature* **303**: 378–9.

Nermut, M.V. and Steven, A.C. (Eds) (1978). Animal virus structure. In *Perspectives in Medical Virology*, Vol. 3. Elsevier, Amsterdam.

Palmer, E.L. and Martin, M.L. (1982). *An Atlas of Mammalian Viruses*. CRC Press, Boca Raton, FL.

Bibliography

Cheng, R.H., Olson, N.H. and Baker, T.S. (1992). Cauliflower mosaic virus: A 420 Subunit (T = 7) multilayer structure. *Virology* **186**: 655–68.

Hogle, J.M. *et al.* (1985). Three-dimensional structure of poliovirus at 2.9 Å resolution. *Science* **229**: 1358–65.

McKenna, R. *et al.* (1992). Atomic structure of single-stranded DNA bacteriophage φX174 and its functional implications. *Nature* **355**: 137–43.

Rasched, I. and Oberer, E. (1986). Ff Coliphages: structural and functional relationships. *Microbiol. Rev.* **50**: 401–27.

Rohrmann, G.F. (1992). Baculovirus structural proteins. *J. Gen. Virol.* **73**: 749–61.

Rossman, M.G. *et al.* (1985). Structure of a human common cold virus and functional relationship to other picornaviruses. *Nature* **317**: 145–53.

Venkataram Prasad, B.V. *et al.* (1992). Three-dimensional structure of single-shelled bluetongue virus. *J. Virol.* **66**: 2135–42.

Chapter 3

Genomes

The Structure and Complexity of Virus Genomes

The compositions and structures of virus genomes are more varied than any of those seen in the entire bacterial, plant or animal kingdoms. The nucleic acid comprising the genome may be single-stranded or double-stranded, and in a linear, circular or segmented configuration. Single-stranded virus genomes may be either **positive (+)sense**, i.e. of the same polarity (nucleotide sequence) as mRNA, **negative (−)sense**, or **ambisense** – a mixture of the two. Virus genomes range in size from approximately 3500 nucleotides (**nt**) (e.g. bacteriophages of the family *Leviviridae* such as MS2 and Qβ) to approximately 280 kilobase pairs (**kbp**), i.e. 560 000 nt (e.g. herpesviruses such as *Cytomegalovirus*). Unlike the genomes of *all* cells, which are composed of DNA, virus genomes may contain their genetic information encoded in either DNA or RNA.

Whatever the particular composition of a virus genome, all must conform to one condition. Since viruses are obligate intracellular parasites only able to replicate inside the appropriate host cells, the genome must contain information encoded in a form which can be recognized and decoded by the particular type of cell parasitized. Thus, the genetic code employed by the virus must match or at least be recognized by the host organism. Similarly, the control signals which direct the expression of virus genes must be appropriate to the host. Chapter 4 describes the means by which virus genomes are replicated and Chapter 5 deals in more detail with the mechanisms that regulate the expression of virus genetic information. The purpose of this chapter is to describe the diversity of virus genomes and to examine how and why this variation may have arisen.

Although molecular biology has developed many techniques for the manipulation of proteins, the power of this technology has concentrated in the main on nucleic acids. There has been a strong and synergistic relationship between advances in virology dependent on this new technology and the opportunities which viruses themselves have afforded to develop new techniques of investigation. Thus, the first complete genome to be sequenced (in the 1970s) was that of a virus, the bacteriophage φX174. This virus was chosen for a number of reasons. First, it has one of the smallest genomes known (5386 nt). Second, large amounts of the phage could be propagated in *E. coli*, easily purified and the genomic DNA extracted. Third, the genome of this phage consists of single-stranded DNA which could be directly sequenced by chain-termination methods. Although methods of sequencing double-stranded DNA were also being developed at this time, ϕX174 proved the utility of bacteriophages with single-stranded genomes as cloning vectors for DNA sequencing and phages such as M13 have subsequently been highly developed for this purpose.

Virus genome structures and nucleotide sequences have been intensively studied in recent years because the power of recombinant DNA technology has focused much attention on the virus genome. It would be wrong to present molecular biology as the only means of addressing unanswered problems in virology, but it would be equally foolish to ignore the opportunities which it offers and the explosion of knowledge which has resulted from it in the last two decades.

Some of the simpler bacteriophages have been cited above as examples of the smallest and least complex genomes known. At the other end of the scale, the genomes of the largest double-stranded DNA viruses such as herpesviruses and poxviruses are sufficiently complex to have escaped complete functional analysis to date (even though the complete nucleotide sequences of the genomes of a growing number of examples are now known). Many of the DNA viruses of eukaryotes closely resemble their host cells in terms of the biology of their genomes:

- Some DNA virus genomes are complexed with cellular histones to form a **chromatin**-like structure inside the virus particle. Once inside the nucleus of the host cell, these genomes behave to all intents and purposes like miniature satellite chromosomes, following the dictates of cellular enzymes and the cell cycle.
- Vaccinia virus mRNAs were found to be polyadenylated at their 3′ ends by J. Kates in 1970 – the first observation of this phenomenon.
- Split genes containing non-coding **introns**, protein coding **exons** and spliced mRNAs were first discovered in adenoviruses by Sharp in 1977.

Introns in prokaryotes were first discovered in the genome of bacteriophage T4 in 1984. Several examples of this phenomenon have now been discovered in T4 and in other phages. All are similar, being of the class I 'self-splicing' type. However, this observation raises an important point. The conventional view is that prokaryote genomes are smaller and replicate faster than those of eukaryotes, and hence can be regarded as 'streamlined.' The genome of phage T4 consists of 160 kbp of double-stranded DNA and is highly compressed; for example, promoters and translation control sequences are nested within the coding regions of overlapping upstream genes. The presence of introns in bacteriophage genomes, which are under constant ruthless pressure to exclude 'junk sequences,' indicates that these genetic elements must have evolved mechanisms to escape or neutralize this pressure and to persist as parasites within parasites.

To a great extent, the same pressure to minimize size exists on all virus genomes. For example, viruses with prokaryotic hosts must be able to replicate sufficiently fast to keep up with their host cells and this is reflected in the compact nature of many (but not all) bacteriophages. Overlapping genes are common and in this way the maximum coding information is compressed into the minimum genome size. In viruses with eukaryotic hosts there is also pressure on genome size. Here, however, the pressure is mainly from the 'packaging size' of the virus particle, i.e. the amount of nucleic acid which can be incorporated into the virion. Therefore, these viruses also commonly show tremendous compression of genetic information when compared with the apparently wasteful low density of information in the genomes of eukaryotic cells.

There are, as ever, exceptions. Some bacteriophages (e.g. the family *Myoviridae*, such as T4) have relatively large genomes, up to 160 kbp. Among viruses of eukaryotes, herpesviruses and poxviruses also have large genomes, up to 280 kbp. It is notable that in these cases, the virus genome contains many genes concerned with its own replication, particularly enzymes concerned with nucleic acid metabolism. Therefore, these viruses appear to be attempting to escape the restrictions of the biochemistry of the host cell by encoding additional biochemical apparatus. The penalty is that they have to encode all the information necessary for a large and complex particle to package the genome – also an upward pressure on genome size.

Later sections of this chapter contain detailed descriptions of both small, compact and large, complex virus genomes.

Molecular Genetics

As already described, the new techniques of molecular biology have had a major influence in concentrating much attention on the virus genome. It is beyond the scope of this book to give detailed accounts of this technology. Indeed, it is assumed that readers already have a firm grasp of the principles behind these techniques, as well as the jargon involved! However, it is perhaps worth taking some time here to illustrate how some of these techniques have been applied to virology, remembering that these newer techniques are complementary to and do not replace the classical techniques of virology. Initially, any investigation of a virus genome will usually include questions about the following:

- Composition – DNA or RNA, single-stranded or double-stranded, linear, or circular.
- Size and number of segments.
- Terminal structures.
- Nucleotide sequence.
- Coding capacity – open reading frames.
- Regulatory signals – transcription enhancers, promoters, and terminators.

It is possible to separate the molecular analysis of virus genomes into two types of approach: physical analysis of structure and nucleotide sequence, essentially performed *in vitro*, and a more biological approach to examine the structure–function relationships of intact virus genomes and individual genetic elements, usually involving analysis of the virus phenotype *in vivo*.

The conventional starting point for the physical analysis of virus genomes has been the isolation of nucleic acids from virus preparations of varying degrees of purity. To some extent, this is still true of molecular biology, although the emphasis on extensive purification has declined as techniques of molecular cloning have become more advanced. DNA virus genomes can be analyzed directly by restriction endonuclease digestion without resort to molecular cloning and this was achieved for the first time with SV40 DNA in 1971. The enzymology of RNA-specific nucleases was comparatively advanced at this time, such that a spectrum of enzymes with known, specific cleavage sites could be used to analyze and even determine the sequence RNA virus genomes (the first short nucleotide sequences of tRNAs having been determined in the mid-1960s). However, direct analysis of RNA by these methods was laborious and notoriously difficult. RNA sequence analysis did not begin to advance rapidly until the widespread use of reverse transcriptase (isolated from avian myeloblastosis virus) to convert RNA into cDNA in the 1970s. The first pieces of DNA to be molecularly cloned were restriction fragments of bacteriophage λ DNA which were cloned into the DNA genome of SV40 by Berg and colleagues in 1972. Thus virus genomes were both the first cloning vectors and the first nucleic acids to be analyzed by these techniques! As already mentioned, the genome of φX174 was the first **replicon** to be completely sequenced and subsequently, phage genomes such as M13 were highly modified for use as vectors in DNA sequencing.

Aside from molecular cloning, other techniques of molecular analysis have also been of great value in virology. Direct analysis by electron microscopy, if calibrated with known standards, can be used to estimate the size of nucleic acid molecules. Perhaps the most important single technique has been gel electrophoresis (Figure 3.1). The earliest gel matrix employed for separating molecules was based on starch and gave relatively poor resolution. Now it is most common to use agarose gels to separate large nucleic acid molecules (which may be very large indeed – several megabases – in the case of techniques such as pulsed-field gel electrophoresis (PFGE)) and polyacrylamide gel electrophoresis (PAGE) to separate smaller pieces (down to the size of very short oligonucleotides). Apart from the fact that nucleotide sequencing is dependent on the ability to separate molecules that differ from each other by only one nucleotide in size, gel electrophoresis has been of great value in analyzing intact virus genomes, and particularly the analysis of viruses with segmented genomes (see below). Hybridization of complementary nucleotide sequences can also be used in a number of ways to analyze virus genomes.

Phenotypic analysis of virus populations has long been a standard technique of virology. Examination of variant viruses and naturally occurring spontaneous mutants has been a longstanding method of determining the function of virus genes. Molecular biology has added to this the ability to design and create specific mutations, deletions and recombinants *in vitro*, a very powerful tool indeed. Although coding capacity and to some extent protein properties can be determined *in vitro* by the use of cell-free extracts to translate mRNAs, complete functional analysis of virus genomes can only be performed on intact viruses. Fortunately, the relative simplicity of virus genomes (compared with even the simplest cell) offers a major advantage here – the ability to 'rescue' infectious virus from purified or cloned nucleic acids. Infection of cells caused by nucleic acid alone is referred to as **transfection**.

Virus genomes which consist of (+)sense RNA are infectious when the purified RNA (vRNA) is applied to cells in the absence of any virus proteins. This is because (+)sense vRNA is essentially mRNA and the first event in a normally infected cell is to translate the vRNA to make the virus proteins responsible for genome

Mixture of
nucleic acids

Molecules move through
gel towards the anode
(+ve terminal) and are
separated by the gel
matrix based on their
size

Figure 3.1 In gel electrophoresis, a mixture of nucleic acids (or proteins) are applied to a gel and move through the gel matrix when an electric field is applied. The net negative charge due to the phosphate groups in the backbone of nucleic acid molecules results in their movement away from the cathode and towards the anode. Smaller molecules are able to slip though the gel matrix more easily and thus migrate further than larger molecules, which are retarded, resulting in a net separation based on the size of the molecules.

replication. In this case, direct introduction of RNA into cells merely circumvents the earliest stages of the replicative cycle (Chapter 4). In most cases, virus genomes which are composed of double-stranded DNA are also infectious. The events which occur here are a little more complex, since the virus genome must first be transcribed by host polymerases to produce mRNA. This is relatively simple for phage genomes introduced into prokaryotes, but for viruses which replicate in the nucleus of eukaryotic cells, such as herpesviruses, the DNA must first find its way to the appropriate cellular compartment. Most of the DNA which is introduced into cells by transfection is degraded by cellular nucleases. However, irrespective of its sequence, a small proportion of the newly introduced DNA finds its way into the nucleus, where it is transcribed by cellular polymerases. Unexpectedly, cloned cDNA genomes of (+)sense RNA viruses (e.g. picornaviruses), are also infectious, although less efficient at infecting cells than the vRNA. This is presumably because the DNA is transcribed by cellular enzymes to make RNA. Using these techniques, virus can be rescued from cloned genomes, including those which have been manipulated *in vitro*.

Until very recently, this type of approach was not open for analysis of viruses with (−)sense genomes. This is because such virus particles all contain a virus-specific polymerase. The first event when the virus genome enters the cell is that the (−)sense genome is copied by the polymerase, forming either (+)sense transcripts which are used directly as mRNA, or a double-stranded molecule known either as the replicative intermediate (RI) or replicative form (RF), which serves as a template for further rounds of mRNA synthesis. Therefore, since purified (−)sense genomes cannot be directly translated and are not replicated in the absence of the virus polymerase, these genomes are inherently non-infectious. Systems have recently been developed which permit the rescue of viruses with (−)sense genomes from purified or cloned nucleic acids. This has been achieved by adding nucleic acid to extracts made from purified virus or virus-infected cells, or by transfecting into 'helper cells' which can supply the required function *in* **trans**. Such developments will open up possibilities for genetic investigation of negative-strand viruses that have not existed before.

It would also be appropriate to mention once again the reliance of molecular biologists on computers to store and process the mass of information which is rapidly generated by techniques of molecular analysis such as nucleotide sequencing. Moreover, as computer software becomes more sophisticated, computers are being used increasingly to make predictions based on nucleotide sequences (see Chapter 1). These include detecting the presence of open reading frames, the amino acid sequences of the proteins encoded by them, control regions of genes such as promoters and splice signals and the secondary structure of proteins and nucleic acids. However (particularly in the case of RNA), it is the secondary structure assumed by molecules that is as important as the primary nucleotide sequence in determining the biological reactions which the molecule may undergo. Much caution is needed in interpreting such predicted rather than factual information, and the validity of such predictions should not be accepted without question unless confirmed by biochemical and/or genetic data.

Virus Genetics

Conventional genetic analysis of animal viruses is based largely on isolation and analysis of mutants. Most commonly, this is achieved using plaque-purification techniques. In the case of viruses for which no such systems exist (either because they are not cytopathic or do not replicate in culture), little genetic analysis was possible before the development of molecular genetics. Nowadays, however, a few tricks make it possible to extend these genetic techniques to non-cytopathic viruses, for example:

- *Transformed foci*: Production of transformed 'foci' of cells by non-cytopathic 'focus-forming' viruses (e.g. DNA and RNA tumor viruses).
- *Focal immunoassays*: replication of non-cytopathic viruses can be visualized by immune-staining to produce visual foci (e.g. human immunodeficiency virus).

- *Biochemical analysis*: the use of metabolic inhibitors to construct genetic 'maps.' Pactamycin, an antibiotic which inhibits initiation of translation, has been used to map protein-coding regions of enteroviruses. Inhibitors of translation (such as puromycin and cycloheximide) and transcription (actinomycin D) can be used to decipher genetic regulatory mechanisms.
- *Physical analysis*: the use of high-resolution electrophoresis to identify genetic polymorphisms of viral proteins or nucleic acids (above).
- *Molecular biology*.

Various types of genetic map can be derived:

- *Recombination maps*: These are an ordered sequence of mutations which can be derived from the probability of recombination between two genetic markers, which is proportional to the distance between them – a classical genetic technique. This method works for viruses with non-segmented genomes (DNA or RNA).
- *Reassortment maps (or groups)*: In viruses with segmented genomes, the assignment of mutations to particular genome segments results in the identification of genetically linked reassortment groups equivalent to individual genome segments.

Other types of map which can be constructed include the following:

- *Physical maps*: Mutations or other features can be assigned to physical locations on a virus genome using the 'rescue' of mutant genomes by small pieces of the wild-type genome (e.g. heteroduplex formation between mutant and wild-type DNA), after transfection of susceptible cells. Alternatively, cells can be co-transfected with the mutant genome plus individual restriction fragments to localize the mutation. Similarly, various polymorphisms (such as electrophoretic mobility of proteins) can be used to achieve the same end.
- *Restriction maps*: Site-specific cleavage of DNA by restriction endonucleases can be used to determine the structure of virus genomes. RNA genomes can be analyzed in this way after cDNA cloning.
- *Transcription maps*: Maps of regions encoding various mRNAs can be determined as follows:
  - Hybridization of mRNA species to specific genome fragments (e.g. restriction fragments). The precise start/finish of mRNAs can be determined by S1 nuclease digestion of radiolabeled probes. Proteins encoded by individual mRNAs can be determined by *in vitro* translation.
  - Ultraviolet (UV) irradiation of RNA virus genomes can be used to determine the position of open reading frames, since those farthest from the translation start are the least likely to be expressed by *in vitro* translation after partial degradation of the virus RNA by UV light.
- *Translation maps*: Pactamycin (which inhibits the initiation of translation) has been used to map protein-coding regions of enteroviruses. Pulse labeling results in incorporation of radioactivity only into proteins initiated prior to addition of the drug. Proteins nearest the 3' end of the genome are the most heavily labeled, those at the 5' end of the genome the least.

Virus mutants

'Strain,' 'type,' 'variant,' 'mutant' and even 'isolate' are all terms used interchangeably to differentiate them from original 'parental,' 'wild-type' or 'street' viruses. More accurately, these terms should be applied as follows:

- *Strain*: Different lines or isolates of the same virus.
- *Type*: Different serotypes of the same virus (e.g. various neutralization phenotypes).
- *Variant*: A virus which differs from the original wild-type strain but where the genetic basis for the difference is not known.

Types of mutant viruses

Spontaneous mutations   In some viruses, mutation rates may be as high as 10^{-3}–10^{-4} per incorporated nucleotide (e.g. in retroviruses such as HIV), whereas in others, it may be as low as 10^{-8}–10^{-11} (e.g. in herpesviruses) – equivalent to the low mutation rates seen in cellular DNA. These differences are due to the mechanism of genome replication, error rates in RNA-dependent RNA polymerases generally being higher than DNA-dependent DNA polymerases. Some RNA virus polymerases do have proofreading functions, but in general, mutation rates are considerably higher in most RNA viruses than in DNA viruses. For a virus, mutations are a mixed blessing. The ability to generate antigenic variants which can escape the immune response is a clear advantage, but mutation also results in many defective particles, since most mutations are deleterious. In the most extreme cases, e.g. HIV, the error rate is 10^{-3}–10^{-4} per nucleotide incorporated. The HIV genome is approximately 9.7 kb long, therefore, there will be 0.9–9.7 mutations in every genome copied. Hence in this case, the 'wild-type' virus actually consists of a fleeting majority type which dominates the dynamic equilibrium (i.e. the population of genomes) present in all cultures of the virus. These mixtures of molecular variants are known as **quasispecies** and also occur in other RNA viruses, e.g. picornaviruses. However, the majority of these types will be non-infectious or seriously disadvantaged and are therefore rapidly weeded-out of a replicating population.

Induced mutations   Most genetic analysis of viruses has been performed on viral mutants isolated from mutagen-treated populations. Mutagenic agents can be divided into two types: *in vitro* mutagens and *in vivo* mutagens.

In vitro mutagens do not require nucleic acid replication for their activity. Examples include:

- nitrous acid, which deaminates bases resulting in AT to GC and CG to TA transitions;
- hydroxylamine, which causes CG to TA transitions; and
- alkylating agents (e.g. nitrosoguanidine), which alkylate the N-7 position of purine nucleotides.

In vivo mutagens require metabolically active, i.e. replicating, nucleic acid for their activity. Examples include:

- base analogs (e.g. 2-aminopurine, 5-bromodeoxyuridine, 5-azacytidine), which cause **transitions** (i.e. purine–purine and pyrimidine–pyrimidine: AT–GC and CG–TA) and **transversions** (i.e. purine–pyrimidine and pyrimidine–purine: AT–TA and GC–CG) due to faulty base-pairing when they are incorporated into nucleic acid. 5-Fluorouridine, an analogue of uracil, is a useful compound for mutating RNA genomes;
- intercalating agents (e.g. acridine dyes), which stack between bases, causing insertions or deletions; and
- UV irradiation, which causes the formation of pyrimidine dimers which are excised from DNA by repair mechanisms that are much more error-prone than the usual enzymes used in DNA replication. This technique also works on RNA, although the mechanism of action is not known in this case.

The main problem with induced mutants is the number of changes per genome. The dose of mutagen used must be chosen carefully to give an *average* (based on Poisson distribution) of < 1 mutation per genome, otherwise the resultant viruses will contain multiple mutations which can complicate interpretation of the phenotype.

Types of mutation

Plaque morphology Mutants may be either large-plaque mutants which replicate more rapidly than the wild-type, or small-plaque mutants, which are the opposite. Plaque size is often related to a temperature-sensitive (t.s.) phenotype (see below). Such mutants are often useful as unselected markers in multifactorial crosses.

Host range This term can refer either to whole animal hosts or to permissive cell types *in vitro*. Conditional mutants of this class have been isolated using amber-suppressor cells (mostly for phages, but also for animal viruses using *in vitro* systems).

Temperature-sensitive (t.s.) These are really 'heat-sensitive' mutants. This type of mutation is very useful as it allows the isolation of conditional-lethal mutations, a powerful means of examining essential virus genes. Such mutations usually result from mis-sense mutations in proteins (i.e. amino acid substitutions), resulting in proteins of full size with subtly altered conformation which can function at (permissive) low temperatures but not at (non-permissive) higher ones. Generally, the mutant proteins are immunologically unaltered, which is frequently a useful attribute. These mutations are usually 'leaky,' i.e. some of the normal activity is retained even at non-permissive temperatures. Conversely, protein function is often impaired, even at the permissive temperature. Therefore, a high frequency of reversion is often a problem with this type of mutation, because the wild-type virus replicates faster than the mutant. In some viruses (e.g. reoviruses, influenza virus), many t.s. mutants have been derived over the years for every virus gene, which has permitted a complete genetic analysis of the genome.

Cold-sensitive (c.s.) These mutants are the opposite of the above class. Cold-sensitive mutants are very useful in bacteriophages and plant viruses as their host cells can be propagated at low temperatures, but are not so valuable for animal viruses because their host cells generally will not grow at lower temperatures than normal. It is possible that some c.s. mutants of respiratory viruses, e.g. influenza virus, may be attenuated (i.e. have reduced virulence) and therefore of interest as potential vaccine strains. This is because influenza virus replicates in the cells of the upper respiratory tract, which are at a slightly lower temperature than those at the body core. Attractive though this idea is, it has not proved possible to demonstrate this phenomenon consistently.

Nonsense   These mutants are also called amber mutations. They may result from alteration of coding sequence to one of three translation stop codons (UAG – *amber*, UAA – *ochre*; UGA – *opal*). Translation is terminated, resulting in the production of an amino-terminal fragment of the protein. The phenotype of these mutations can be suppressed by propagation of virus in a cell (bacterial or, more recently, animal) with altered suppressor tRNAs. Unlike t.s. mutants, nonsense mutations are rarely leaky and can only revert to wild-type at the original site (see below). They therefore usually show a low reversion frequency.

Deletions   These are similar in some ways to nonsense mutants, but may include one or more virus genes and also involve non-coding control regions (promoters etc.). Spontaneous deletion mutants often accumulate in virus populations as 'defective-interfering' (DI) particles. These non-infectious but not necessarily genetically inert genomes are thought to be important in establishing the course and pathogenesis of certain virus infections (see Chapter 6). Genetic deletions can only revert to wild-type by recombination, which usually occurs at comparatively low frequencies. Deletion mutants are very useful for assigning structure–function relationships to virus genomes, since they are easily mapped by physical analysis.

Biochemical markers   This category includes drug resistance mutations; specific mutations which result in altered virulence; polymorphisms resulting in altered electrophoretic mobility of proteins or nucleic acids; and altered sensitivity to inactivating agents.

Revertants   Reverse mutation is a valid type of mutation in its own right. Most of the above classes can undergo reverse mutations, which may be either direct 'back mutations' (i.e. correction of the original mutation) or second site 'compensatory mutations,' which may be physically distant from the original mutation, and not even necessarily in the same gene as the original mutation.

Genetic interactions between viruses

Genetic interactions between viruses often occur naturally *in vivo*, since host organisms are frequently infected with more than one virus. However, these situations are generally much too complicated to analyze successfully. Experimentally, genetic interactions can be analyzed by mixed infection (**superinfection**) of cells in culture. Two types of information can be obtained from such experiments:

- The assignment of mutants to functional groups known as complementation groups (below).
- The ordering of mutants into a linear genetic map by analysis of recombination frequencies (i.e. recombination groups).

Complementation can be defined as an interaction of virus gene products in infected cells that results in the yield of one or both of the parental mutants being enhanced while their genotypes remain unchanged. In this situation, one of the viruses in a mixed infection provides a functional gene product for another virus which is defective for that function (Figure 3.2). If both mutants are defective in the same function, enhancement of replication does not occur and the two mutants are said to be in the same complementation group. The importance of this test is that it allows functional analysis of unknown mutations, assuming that the biochemical basis of any one of the mutations in a particular complementation group is known. In theory, the number of complementation groups is equal to the number of genes in the virus genome. In practice, there are usually fewer complementation groups than genes, since mutations in some genes are always lethal and other genes are non-essential and therefore cannot be scored in this type of test. There are two possible types of complementation:

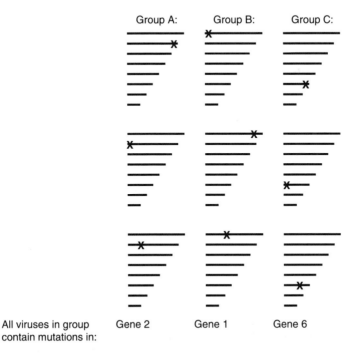

Figure 3.2 Complementation groups of influenza (or other) viruses all contain a mutation in the same virus gene, preventing the rescue of another mutant virus genome from the same complementation group.

- Allelic (intragenic) complementation occurs where different mutants have complementing defects in the same protein, e.g. in different functional domains, or in different subunits of a multimeric protein (although this is rare).
- Non-allelic (intergenic) complementation results from mutants with defects in different genes. This is the more common type.

Complementation can be asymmetric, i.e. only works in one direction and, therefore, only one of the parental viruses will be represented among the progeny. This can be an absolute or a partial restriction. When complementation occurs naturally, it is usually the case that a replication-competent wild-type virus 'rescues' a replication-defective mutant. In these cases, the wild-type is referred to as a 'helper virus,' for example, as in the case of defective transforming retroviruses containing oncogenes (whose genomes are deleted because of packaging constraints) (Figure 3.3).

Recombination is the physical interaction of virus genomes in a mixed infection resulting in progeny genomes containing genetic information in non-parental combinations. There are three mechanisms by which this can occur, depending on the organization of the virus genome:

(1) *Intramolecular recombination by strand breakage and re-ligation*: This process occurs in all DNA viruses and RNA viruses which replicate via a DNA intermediate. It is believed to be mediated by cellular enzymes, since no virus mutants with specific recombination defects have been isolated.

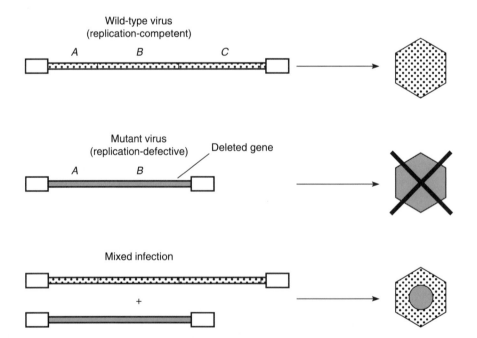

Figure 3.3 Helper viruses are replication-competent viruses which are capable of rescuing replication-defective genomes in a mixed infection, permitting their multiplication and spread.

(2) *Intramolecular recombination by 'copy-choice:'* This process occurs in RNA viruses (it has been known in picornaviruses since the 1960s, but has been recognized in other virus groups, such as coronaviruses, only recently), probably by a mechanism in which the virus polymerase switches template strands during genome synthesis. The molecular details of this process are not well understood. There are cellular enzymes which could be involved (e.g. splicing enzymes), but this is unlikely and the process is thought to occur essentially as a random event. Defective interfering (DI) particles in RNA virus infections are frequently generated in this way (see Chapter 6).

(3) *Reassortment*: In viruses with segmented genomes, the genome segments can be randomly shuffled during a mixed infection. Progeny viruses receive one of each of the genome segments, but probably not from a single parent. For example, influenza virus has eight genome segments, therefore in a mixed infection, there could be $2^8 = 256$ possible progeny viruses. Packaging mechanisms in these viruses are not understood but may be involved in generating reassortants.

In intramolecular recombination, the probability that breakage-reunion or strand-switching will occur between two markers (resulting in recombination) is proportional to the physical distance between them. Therefore, pairs of markers will show a range of recombination frequencies up to a (theoretical) maximum of 50% and can be arranged on a linear genetic map, with distances measured in 'map units,' i.e. percentage recombination frequency. In reassortment, the frequency of recombinants between two markers is either very high (indicating that the markers are on two different genome segments) or comparatively low (which means that they are on the same segment). The frequency of reassortment usually swamps the lower background frequency which is due to intramolecular recombination within strands.

Reactivation is the generation of infectious (recombinant) progeny from two non-infectious parental virus genomes. This process has been demonstrated *in vitro* and may be important *in vivo*; for example, it has been suggested that the rescue of defective, long dormant HIV proviruses during the long clinical course of AIDS may result in increased antigenic diversity and contribute to the pathogenesis of the disease. Recombination has been shown many times to occur in nature, e.g. influenza virus reassortment has resulted in worldwide epidemics (**pandemics**) which have killed millions of people. This makes these genetic interactions of considerable practical interest and not merely a dry academic matter.

Non-genetic interactions between viruses

A number of non-genetic interactions between viruses occur which can affect the outcome and interpretation of the results of genetic crosses. Eukaryotic cells have a diploid genome with two copies of each chromosome, each bearing its own allele of the same gene. The two chromosomes may differ in allelic markers at many loci. Among viruses, only retroviruses are truly diploid with two complete copies of the entire genome, but some DNA viruses, such as herpesviruses, have repeated sequences and are therefore partially heterozygous. In a few (mostly enveloped) viruses, aberrant packaging of multiple genomes may on occasion result in multiploid particles which are heterozygous (e.g. up to 10% of Newcastle disease virus particles). This process is known as **heterozygosis** and can contribute to the genetic complexity of virus populations.

Another non-genetic interaction which is commonly seen between different viruses is **interference**. This process results from the resistance to superinfection by a virus seen in cells already infected by another virus. Homologous interference (i.e. only occurs against the same virus) often results from the presence of DI particles which compete for essential cell components and block replication. However, interference can also result from other types of mutation (e.g. dominant t.s. mutations).

Suppression is the inhibition of a mutant phenotype by a second suppressor mutation, which may be either in the virus genome or in that of the host cell. It is important to point out that this mechanism of suppression is not the same as the suppression of chain-terminating amber mutations by host-encoded suppressor tRNAs (above), which could be called 'informational suppression.' Genetic suppression results in an apparently wild-type phenotype from a virus which is still genetically mutant – a **pseudorevertant**. This phenomenon has been best characterized in prokaryotic systems. More recently, some examples have been discovered in animal viruses, for example, reoviruses, vaccinia, influenza (where it has been observed in an attenuated vaccine, leading to an apparently virulent virus – therefore this could be medically important). Suppression may also be important biologically, in allowing viruses to overcome the deleterious effects of mutations, and would therefore be positively selected. Returning to the issue of reversion already discussed above, mutant viruses can therefore appear to revert by any one of three pathways:

- Back mutation of the original mutation to give a wild-type genotype/phenotype (true reversion).
- A second, compensatory mutation may occur in the same gene as the original mutation, correcting it (e.g. a second frameshift mutation restoring the original reading frame – intragenic suppression).
- A suppressor mutation in a different virus gene or a host gene (extragenic suppression).

Phenotypic mixing is where individual progeny viruses from a mixed infection contain structural proteins derived from both parental viruses. This can vary from extreme cases, where the genome of one virus is completely enclosed within the capsid or envelope of the other, to more subtle cases where the capsid/envelope of the progeny contains a mixture of proteins from both viruses. This mixing gives the progeny virus the phenotypic properties (e.g. cell tropism) dependent on the proteins incorporated into the particle, *without any genetic change*. Subsequent generations of viruses show the original parental phenotypes. This process can occur easily in viruses with naked capsids (non-enveloped) which are closely related (e.g. different strains of enteroviruses), or in enveloped viruses (rescue of replication-defective transforming retroviruses by helper virus is a form of pseudotyping) which need not be related to one another (Figure 3.4). In this latter case, the phenomenon is due to the rather non-specific incorporation of different

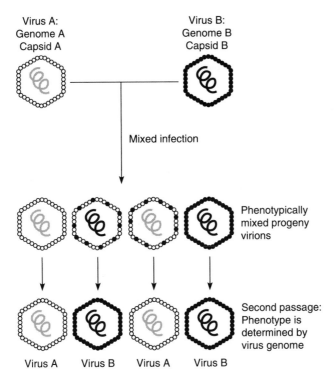

Virus A:
Genome A
Capsid A

Virus B:
Genome B
Capsid B

Mixed infection

Phenotypically
mixed progeny
virions

Second passage:
Phenotype is
determined by
virus genome

Virus A   Virus B   Virus A   Virus B

Figure 3.4 Phenotypic mixing occurs in mixed infections, resulting in genetically unaltered virus particles which have some of the properties of the other parental type.

virus glycoproteins into the envelope, resulting in a mixed phenotype. Phenotypic mixing has proved to be a very useful tool to examine biological properties of viruses. Vesicular stomatitis virus readily forms pseudotypes containing retrovirus envelope glycoproteins, giving a plaque-forming virus with the properties of VSV with the cell tropism of the retrovirus. This trick has been used in recent years to study the cell tropism of HIV and other retroviruses.

'Large' DNA Genomes

There are a number of virus groups which have double-stranded DNA genomes of considerable size and complexity. In many respects, these viruses are genetically very similar to the host cells which they infect. Two examples are the members of the *Adenoviridae* and *Herpesviridae* families.

The *Herpesviridae* are a large family containing more than 100 different members, at least one for most animal species which have been examined to date. There are seven human herpesviruses, all of which share a common overall genome structure, but which differ in the fine details of genome organization and at the level of nucleotide sequence. The family is divided into three subfamilies, based on their biological properties:

Alphaherpesvirinae	Latent infections in sensory ganglia, e.g. herpes simplex virus (HSV), varicella-zoster virus (VZV). Genome size 120–180 kbp.
Betaherpesvirinae	Restricted host range, e.g. cytomegalovirus (CMV), HHV-6. Genome size 180–230 kbp.
Gammaherpesvirinae	Infection of lymphoblastoid cells, e.g. Epstein–Barr virus (EBV). Genome size 105–170 kbp.

Herpesviruses have very large genomes composed of up to 230 kbp of linear, double-stranded DNA and correspondingly large and complex virus particles, containing more than 30 virion polypeptides. All encode a variety of enzymes involved in nucleic acid metabolism, DNA synthesis and protein processing (e.g. protein kinases). The different members of the family are widely separated in terms of genomic sequence and proteins, but all are similar in terms of structure and genome organization (Figure 3.5a). Some but not all herpesvirus genomes consist of two covalently joined sections, a unique long (U_L) and a unique short (U_S) region, each bounded by inverted repeats. The repeats allow structural rearrangements of the unique regions and, therefore, these genomes exist as a

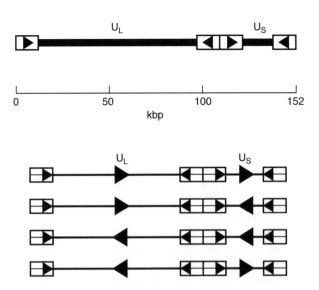

Figure 3.5 (a) Some herpesvirus genomes (e.g. herpes simplex virus) consist of two covalently joined sections, U_L and U_S, each bounded by inverted repeats. (b) This organization permits the formation of four different isomeric forms of the genome.

mixture of four isomers, all of which are functionally equivalent (Figure 3.5b). Herpesvirus genomes also contain multiple repeated sequences and, depending on the number of these, the genome size of various isolates of a particular virus can vary by up to 10 kbp.

The prototype member of the family is herpes simplex virus (HSV), whose genome consists of approximately 152 kbp of double-stranded DNA, the complete nucleotide sequence of which has now been determined. This virus contains at least 75 genes, densely packed and with overlapping reading frames. However, each gene is expressed from its own promoter (cf. adenoviruses – below). A number of other herpesvirus genomes have now been completely sequenced. To date, nucleotide sequences have not been used as a major criterion in classification of herpesviruses, but the advantage of unequivocal demonstration of the degree of relatedness offered by these data means that it will certainly be used in this way in the future. Prior to the advent of nucleotide sequencing, the HSV genome had already been extensively mapped by 'conventional' genetic analysis, including the study of a very large number of t.s. mutants in many genes. Indeed, HSV is perhaps the most intensively studied of this type of complex virus genome.

In contrast to herpesviruses, the genomes of adenoviruses consist of linear, double-stranded DNA of 30–38 kbp, the precise size of which varies between groups. These virus genomes contain 30–40 genes (Figure 3.6). The terminal sequence of each DNA strand is an inverted repeat of 100–140 bp and, therefore, the denatured single strands can form 'pan-handle' structures. These structures are important in DNA replication, as is a 55 kDa protein known as the terminal protein covalently attached to the 5' end of each strand. During genome replication, this protein acts as a primer, initiating the synthesis of new DNA strands. Although adenovirus genomes are considerably smaller than those of herpesviruses, the expression of the genetic information is rather more complex. Clusters of genes are expressed from a limited number of shared promoters. Multiply spliced mRNAs and alternative splicing patterns are used to express a variety of polypeptides from each promoter (see Chapter 5).

Human adenoviruses are divided into six groups (subgenera) and in this case, genome structure (determined by cross-hybridization and restriction enzyme mapping) is an additional character which has been used to assign viruses to groups. There is 70–95% nucleotide sequence similarity across the entire genome within groups, 5–20% similarity between different groups.

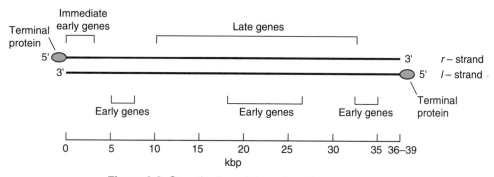

Figure 3.6 Organization of the adenovirus genome.

'Small' DNA Genomes

The structure of the bacteriophage M13 genome has already been described in Chapter 2. Since unlike other virion structures, the filamentous M13 capsid can be expanded by the addition of further protein subunits, the genome size can also be increased by the addition of extra sequences in the non-essential intergenic region without the penalty of becoming incapable of being packaged into the capsid. In other bacteriophages, the packaging constraints are much more rigid; for example, in phage λ, only DNA of between approximately 95 and 110% (approximately 46–54 kbp) of the normal genome size (49 kbp) can be packaged into the virus particle. Not all bacteriophages have such simple genomes as M13, for example, the genome of phage λ is approximately 49 kbp and that of phage T4 is about 160 kbp double-stranded DNA. These latter two bacteriophages also illustrate another common feature of linear virus genomes – the importance of the sequences present at the ends of the genome.

In the case of phage λ, the substrate which is packaged into the phage heads during assembly consists of long concatemers of phage DNA which are produced during the later stages of vegetative replication. The DNA is apparently 'reeled in' by the phage head and when a complete genome has been incorporated, the DNA is cleaved at a specific sequence by a phage-coded endonuclease (Figure 3.7). This enzymes leaves a 12 bp 5′ overhang on the end of each of the cleaved strands, known as the *cos* site. Hydrogen bond formation between these 'sticky ends' can result in the formation of a circular molecule. In a newly infected cell, the gaps on either side of the *cos* site are closed by DNA ligase and it is this circular DNA which it undergoes vegetative replication or integration into the bacterial chromosome.

Bacteriophage T4 and other phages of *E. coli* illustrate another molecular feature of certain linear virus genomes, **terminal redundancy**. Replication of the T4 genome also produces long concatemers of DNA. These are cleaved by a specific endonuclease, but unlike the λ genome, the lengths of DNA incorporated into the particle are somewhat longer than a complete genome length (Figure 3.8). Therefore, some genes are repeated at each end of the genome, and the DNA packaged into the phage particles contains reiterated information. Bacteriophage genomes are neither necessarily small nor simple!

As examples of small DNA genomes, consider those of two groups of animal viruses, the parvoviruses and polyomaviruses. The *Parvoviridae* family contains three genera:

Parvovirus	Autonomous (replication-competent) pathogenic mammalian viruses.
Dependovirus	Replication-defective adenovirus-associated viruses (AAV).
Densovirus	Autonomous insect viruses.

Parvovirus genomes are linear, non-segmented, single-stranded DNA of about 5 kb. Most of the strands packaged into virions are (−)sense, but AAVs package equal amounts of (+) and (−) strands, and all seem to package at least a proportion of (+)sense strands. These are very small genomes, and even the replication-competent parvoviruses contain only two genes: *rep*, which encodes proteins involved in

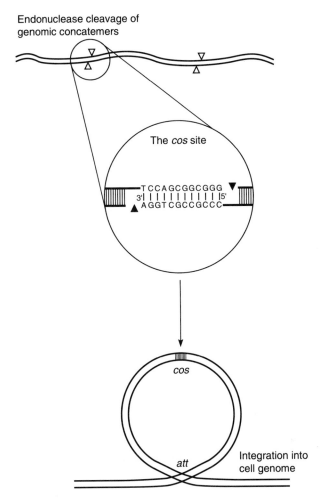

Endonuclease cleavage of
genomic concatemers

The *cos* site

—TCCAGCGGCGGG ▼
3'| | | | | | | | | | | | |5'
▲AGGTCGCCGCCC—

cos

att

Integration into
cell genome

Figure 3.7 The cohesive 'sticky-ends' of the *cos* site in the bacteriophage λ genome are ligated together in newly infected cells, forming a circular molecule. Integration of this circular form into the *E. coli* chromosome occurs by specific recognition and cleavage of the *att* site in the phage genome.

transcription and *cap*, which encodes the coat proteins. However, the expression of these genes is rather complex, resembling the pattern seen in adenoviruses, with multiple splicing patterns seen for each gene (Chapter 5). The ends of the genome have palindromic sequences of about 115 nt, which form 'hairpins' (Figure 3.9). These structures are essential for the initiation of genome replication, once again emphasizing the importance of the sequences at the ends of the genome.

The classification of parvoviruses into genera is made on the functional capabilities of the virus genome. *Dependovirus* was the first group to be discovered. This genus contains four types of adeno-associated viruses (AAV1–4) which are entirely dependent on adenoviruses (or herpesviruses) for essential helper functions required for their replication. The adenovirus genes necessary for *Dependovirus* replication

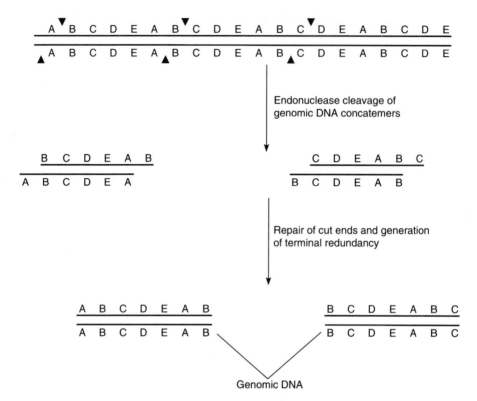

Figure 3.8 Terminal redundancy in the bacteriophage T4 genome results in the reiteration of some genetic information.

Figure 3.9 Palindromic sequences at the ends of parvovirus genomes result in the formation of hairpin structures involved in the initiation of replication.

encode early (regulatory) proteins rather than late (structural) adenovirus proteins, but it has recently been shown that treatment of cells with UV light, cycloheximide, or some carcinogens can replace the requirement for helper virus. Therefore, the *Dependovirus* requirement appears to be for a modification of the cellular environment rather than a specific virus protein – probably affecting transcription of the defective virus genome.

The genomes of polyomaviruses consist of double-stranded, circular DNA molecules, approximately 5 kbp in size. The entire nucleotide sequence of all the viruses in the genus is known and the architecture of the polyomavirus genome (i.e. number and arrangement of genes and function of the regulatory signals and systems) has been studied in great detail at a molecular level. Within the particles, the virus DNA assumes a supercoiled form and is associated with four cellular histones H2A, H2B, H3 and H4 (see Chapter 2). The genomic organization of these viruses has evolved to pack the maximum information (six genes) into minimal space (5 kbp). This has been achieved by the use of both strands of the genome DNA and overlapping genes (Figure 3.10). VP1 is encoded by a dedicated open reading frame (**ORF**), but the VP2 and VP3 genes overlap so that VP3 is contained within VP2. The origin of replication is surrounded by non-coding regions which control transcription. Polyomaviruses also encode 'T-antigens' – proteins which can be detected by sera from animals bearing polyomavirus-induced tumors. These proteins bind to the origin of replication and show complex activities, in that they are involved both in DNA replication and in the transcription of virus genes. This is discussed further in Chapter 7.

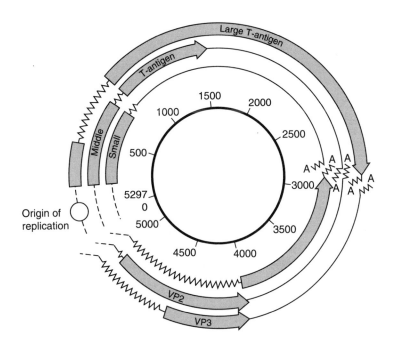

Figure 3.10 The complex organization of the polyomavirus genome results in the compression of much genetic information into a relatively short sequence.

Positive-Strand RNA Viruses

The ultimate size of single-stranded RNA genomes is limited by the fragility of RNA and the tendency of long strands to break. In addition, RNA genomes tend to have higher mutation rates than those composed of DNA because they are copied less accurately. This has also tended to drive RNA viruses towards smaller genomes. Single-stranded RNA genomes vary in size from those of coronaviruses, which are approximately 30 kb long, to those of bacteriophages such as MS2 and Qβ, at about 3.5 kb. This section will consider the features of (+)sense RNA viruses of vertebrates: picornaviruses, togaviruses, flaviviruses, and coronaviruses. Although clearly members of distinct families, all of these viruses share common features in terms of the biology of their genomes: purified (+)sense virus RNA is directly infectious when applied to susceptible host cells in the absence of any virus proteins (although it is about one million times less infectious than virus particles). On examining the features of each of these four families, it is evident that although the details of genomic organization may vary, repeated themes emerge (Figure 3.11).

Picornaviruses

The picornavirus genome consists of one single-stranded, (+)sense RNA molecule of between 7.2 kb in human rhinoviruses (HRV) to 8.5 kb in foot-and-mouth disease viruses (FMDV), containing a number of features conserved in all picornaviruses:

Figure 3.11 Genomic organization of positive-stranded RNA viruses.

- There is a long (600–1200 nt) untranslated region (UTR) at the 5′ end which is important in translation, virulence, and possibly encapsidation and a shorter 3′ untranslated region (50–100 nt) necessary for (−)strand synthesis during replication.
- The rest of the genome encodes a single **polyprotein** of between 2100 and 2400 amino acids.
- Both ends of the genome are modified, the 5′ end by a covalently attached small, basic protein VPg (23 amino acids), the 3′ end by polyadenylation. The polyadenylic acid sequences are not genetically coded – there is a 'polyadenylation signal' upstream of the 3′ end, as in eukaryotic mRNAs.

Togaviruses

The togavirus genome comprises single-stranded, (+)sense, non-segmented RNA of approximately 11.7 kb. It has the following features:

- It resembles cellular mRNAs in that it has a 5′ methylated cap and 3′ poly(A) sequences.
- Expression is achieved by two rounds of translation, producing first non-structural proteins encoded in the 5′ part of the genome and later structural proteins from the 3′ part.

Flaviviruses

The flavivirus genome comprises one single-stranded, (+)sense RNA molecule of about 10.5 kb with the following features:

- It has a 5′ methylated cap but the RNA is not polyadenylated at the 3′ end.
- Genetic organization differs from that of the togaviruses (above) in that the structural proteins are encoded in the 5′ part of genome and non-structural proteins in the 3′ part.
- Expression is similar to that of the picornaviruses, involving the production of a polyprotein.

Coronaviruses

The coronavirus genome consists of non-segmented, single-stranded, (+)sense RNA, approximately 27–30 kb long – the longest of any RNA virus. It also has the following features:

- It has a 5′ methylated cap and 3′ poly(A) and the vRNA functions directly as mRNA.
- The 5′ 20 kb segment of the genome is translated first to produce a viral polymerase, which then produces a full-length (−)sense strand. This is used as a template to produce mRNA as a 'nested set' of transcripts, all with an identical 5′ non-translated leader sequence of 72 nt and coincident 3′ polyadenylated ends.
- Each mRNA is monocistronic, the genes at the 5′ end being translated from the longest mRNA and so on. These unusual cytoplasmic structures are produced

not by splicing (post-transcriptional modification) but by the polymerase during transcription.

The similarities and differences between genomes in this class will be considered further in the discussion of virus evolution below and in Chapter 5.

Negative-Strand RNA Viruses

Viruses with negative-sense RNA genomes are a little more diverse than the positive-stranded viruses discussed above. Possibly because of the difficulties of expression, they tend to have larger genomes encoding more genetic information. Because of this, segmentation is a common, though not universal, feature of such viruses (Figure 3.12). None of these genomes are infectious as purified RNA. Since uninfected cells do not contain RNA-dependent RNA polymerase activity, and since

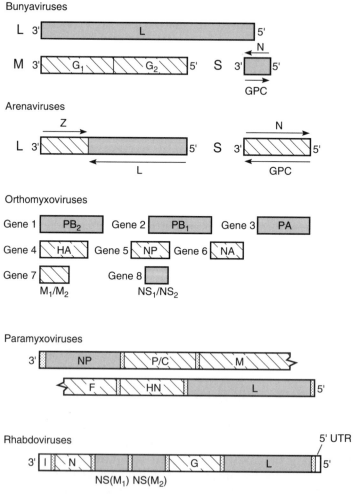

Figure 3.12 Genomic organization of negative-stranded RNA viruses.

the (−)sense genome cannot be translated as mRNA without the virus polymerase packaged in each particle, these genomes are effectively inert. Some of the viruses described in this section are not strictly 'negative-sense' but **ambisense**, since they are part (−)sense and part (+)sense.

Bunyaviruses

Members of the *Bunyaviridae* family have single-stranded, (−)sense, segmented RNA. The genome has the following features:

- The genome comprises three molecules: L (8.5 kb), M (5.7 kb) and S (0.9 kb).
- All three RNA species are linear, but in the virion they appear circular because the ends are held together by base-pairing. The three segments are not present in virus preparations in equimolar amounts.
- In common with all (−)sense RNAs, the 5′ ends are not capped and the 3′ ends are not polyadenylated.
- The members of the *Phlebovirus* genus differ from the other four genera in the family (*Bunyavirus, Nairovirus, Hantavirus,* and *Tospovirus*) in that genome segment S is rather smaller and the overall genome organization is different – ambisense (i.e. the 5′ end of each segment is (−)sense, but the 3′ end is (+)sense).

Arenaviruses

The arenaviruses have linear, single-stranded RNA. The genome organization is ambisense, as in the *Phlebovirus* genus of bunyaviruses. There are two genome segments: L (5.7 kb) and S (2.8 kb).

Orthomyxoviruses

See discussion of segmented genomes (page 70).

Paramyxoviruses

Members of the *Paramyxoviridae* family have non-segmented (−)sense RNA of 17–20 kb. The six genes are organized in a linear arrangement and are separated by repeated sequences: a polyadenylation signal at the end of the gene, an intergenic sequence (GAA), and a translation start signal at the beginning of the next gene.

Rhabdoviruses

Viruses of the *Rhabdoviridae* family have non-segmented, (−)sense RNA of approximately 11 kb. There is a leader region of approximately 50 nt at the 3′ end of the genome and a 60 nt untranslated region (UTR) at the 5′ end of the vRNA. Overall, the genetic arrangement is similar to that of paramyxoviruses, with a conserved polyadenylation signal at the end of each gene and short intergenic regions between the five genes.

Segmented and Multipartite Virus Genomes

There is often some confusion over these two categories of genome structure. Segmented virus genomes are those which are divided into two or more physically separate molecules of nucleic acid, all of which are then packaged into a single virus particle. In contrast, although multipartite genomes also segmented, each genome segment is packaged into a separate virus particle. These discrete particles are structurally similar and may contain the same component proteins, but often differ in size depending on the length of the genome segment packaged. In one sense, multipartite genomes are, of course, segmented, but this is not the strict meaning of these terms as they will be used here. There are many examples of the former strategy, including many human, animal, and plant pathogens such as orthomyxoviruses, reoviruses, and bunyaviruses. However, there are rather fewer examples of multipartite viruses, all of which infect plants. These include bipartite viruses (which have two genome segments and therefore two virus particles) of the *Geminivirus* (single-stranded DNA), *Cryptovirus* (double-stranded RNA), *Comovirus*, *Tobravirus*, and *Furovirus* (single-stranded RNA) groups, and tripartite viruses (three genome segments/virus particles) such as members of the *Bromovirus*, *Cucumovirus*, *Ilarvirus* and *Hordeivirus* (single-stranded RNA) groups.

Segmentation of the virus genome has a number of advantages and disadvantages. There is an upper limit to the size of a non-segmented virus genome which results from the physical properties of nucleic acids, particularly the tendency of long molecules to break due to shear forces (and for each particular virus, the length of nucleic acid which can be packaged into the capsid). The problem of strand breakage is particularly relevant for single-stranded RNA, which is much more chemically labile than double-stranded DNA. The longest single-stranded RNA genomes are those of the coronaviruses, at approximately 30 kb, but the longest double-stranded DNA virus genomes are considerably longer, e.g. cytomegalovirus at approximately 280 kbp. Physical breakage of the genome results in its biological inactivation, since it cannot be completely transcribed, translated, or replicated. Segmentation means that the virus avoids having all its 'eggs in one basket' and also reduces the probability of breakages due to shearing, thus increasing the total potential coding capacity of the entire genome. However, the disadvantage of this strategy is that all the individual genome segments must be packaged into each virus particle, or the virus will be defective as a result of loss of genetic information. In general, it is not understood how this control of packaging is achieved.

Separating the genome segments into different particles (the multipartite strategy) removes the requirement for accurate sorting, but introduces a new problem in that all the discrete virus particles must be taken up by a single host cell to establish a productive infection. This is perhaps the reason why multipartite viruses are only found in plants. Many of the sources of infection by plant viruses, such as inoculation by sap-sucking insects or after physical damage to tissues, result in a large input of infectious virus particles, providing the opportunity for infection of an initial cell by more than one particle.

The genetics of segmented genomes are essentially the same as those of non-segmented genomes, with the addition of the reassortment of segments, as discussed above. Reassortment can occur whether the segments are packaged into a single particle or are in a multipartite configuration. Reassortment is a powerful means of achieving rapid generation of genetic diversity; this could be another possible reason for its evolution. Segmentation of the genome also has implications for the partition of genetic information and the way in which it is expressed; this will be considered further in Chapter 5.

To understand the complexity of these genomes, consider the organization of a segmented virus genome (orthomyxovirus – influenzavirus) and a multipartite genome (geminivirus). The influenzavirus genome is composed of eight segments (in influenza A and B strains, seven in influenza C) of single-stranded, (−)sense RNA, as shown in Table 3.1. The identity of the proteins encoded by each genome segment were determined originally by genetic analysis of the electrophoretic mobility of the individual segments from reassortant viruses and by analysis of a large number of mutants covering all the eight segments. All eight segments have common nucleotide sequences at the 5′ and 3′ ends (Figure 3.13) which are necessary for replication of the genome (Chapter 4). These sequences are complementary to one another and, inside the particle, the ends of the genome segments are held together by base-pairing, forming a 'pan-handle' structure, which again is believed to be involved in replication. The RNA genome segments are not packaged as naked nucleic acid, but in association with the gene 5 product, the nucleoprotein, and are visible in electron micrographs as helical structures. Here, there is a paradox. Biochemically and genetically, each genome segment behaves as an individual, discrete entity. However, in electron micrographs of influenza virus particles disrupted with non-ionic detergents, the nucleocapsid has the physical appearance of a single, long helix. Clearly, there is some interaction between the genome segments and it is this which explains the ability of influenza virus particles to select and package the genome segments within each particle with a surprisingly low error rate, considering the difficulty of the task (Chapter 2). The genome sequences and protein interaction which operate this subtle mechanism have not yet been defined.

The *Geminivirus* group is divided into three taxonomic groups based on their host plants (monocotyledons or dicotyledons) and insect vectors (leafhoppers or

Table 3.1 Segments of the influenzavirus genome

Segment	Size (nt)	Polypeptide(s)	Function
1	2341	PB_2	Polymerase – cap binding
2	2341	PB_1	Polymerase – endonuclease (?)
3	2233	PA	Polymerase – elongation
4	1778	HA	Hemagglutinin
5	1565	NP	Nucleoprotein
6	1413	NA	Neuraminidase
7	1027	M_1	Matrix protein (non-structural)
		M_2	Non-structural – function unknown
8	890	NS_1	Non-structural – function unknown
		NS_2	Non-structural – function unknown

Figure 3.13 Common terminal sequences of influenza virus RNAs.

whiteflies). In groups I and II, the genome consists of a single-stranded DNA molecule of approximately 2.7 kb. The DNA packaged into these virions has been arbitrarily designated as (+)sense, although in fact both the (+) and (−)sense strands found in infected cells contain protein coding sequences. The genome of the group III geminiviruses is bipartite, and consists of two circular, single-stranded DNA molecules, each of which is packaged into a discrete particle (Figure 3.14). Both of the strands comprising the genome are approximately 2.7 kb long and differ from one another completely in nucleotide sequence, except for a shared 200 nt sequence involved in DNA replication. The two genomic DNAs are packaged into entirely separate capsids. Since establishment of a productive infection requires both parts of the genome, it is then necessary for a minimum of two virus particles bearing one copy of each of the genome segments to infect a new host cell. Although geminiviruses do not multiply in the tissues of their insect vectors (**non-propagative** transmission), a sufficiently large amount of virus is ingested and subsequently deposited onto a new host plant to favor such superinfections.

Both of these examples show a high density of coding information. In influenza virus, genes 7 and 8 both encode two proteins in overlapping reading frames. In geminiviruses, both strands of the virus DNA found in infected cells contain coding information, some of which is present in overlapping reading frames. It may be that this high density of genetic information is the reason why these viruses have resorted to divided genomes, in order to regulate the expression of this information (see Chapter 5).

(+) Sense open reading frames shown by clockwise arrows (shaded)
(–) Sense open reading frames shown by anticlockwise arrows (open)

Figure 3.14 Organization and protein-coding potential of the bipartite geminivirus genome.

Reverse Transcription and Transposition

The first successes of molecular biology were the discovery of the double helix structure of DNA and the elucidation of the language of the genetic code. The importance of these findings does not lie in these mere facts, but in their importance in allowing predictions to be made about the fundamental nature of living organisms. The confidence which flowed from these early triumphs resulted in the development of a grand universal theory, which was called the 'central dogma of molecular biology,' namely that all cells (and hence viruses) worked on a simple organizing principle, the unidirectional flow of information from DNA, through RNA, into proteins. In the mid-1960s, there were rumblings that 'life might not be so simple.'

In 1963, Howard Temin showed that the replication of retroviruses, whose particles contain RNA genomes, was inhibited by actinomycin D, an antibiotic which binds only to DNA. The replication of other RNA viruses is not inhibited by this drug. So pleased was the scientific community with its all-embracing dogma that these facts were largely ignored until 1970, when Temin and David Baltimore simultaneously published the observation that retrovirus particles contain an RNA-dependent DNA polymerase – reverse transcriptase. This finding alone was important enough, but like the earlier conclusions of molecular biology, has subsequently had reverberations for the genomes of all organisms, and not merely a few virus families. It is now known that retrotransposons with striking similarities to retrovirus genomes form a substantial part of the genomes of all higher organisms, including man. Earlier ideas of genomes as constant, stable structures have been replaced with the realization that they are, in fact, dynamic and rather fluid entities.

The concept of transposable genetic elements – specific sequences that are able to move from one position in the genome to another – was put forward by Barbara McClintock in the 1940s. Such **transposons** fall into two groups:

- Simple transposons, which do not undergo reverse transcription and are found in prokaryotes (e.g. the genome of bacteriophage Mu).
- Retrotransposons, which closely resemble retrovirus genomes and are bounded by long direct repeats (long terminal repeats – LTRs). These move by means of a transcription/reverse transcription/integration mechanism and are found in eukaryotes (e.g. the yeast Ty element and retrovirus genomes).

Both types show a number of similar properties:

- They are believed to be responsible for a high proportion of apparently 'spontaneous' mutations.
- They promote a wide range of genetic rearrangements in host cell genomes, such as deletions, inversions, duplications, and translocations of the neighboring cellular DNA.
- The mechanism of insertion generates a short (3–13 bp) duplication of the DNA sequence on either side of the inserted element.
- The ends of the transposable element consist of inverted repeats, 2–50 bp in length.
- Transposition is often accompanied by replication of the element – necessarily so in the case of retrotransposons, but this also often occurs with prokaryotic transposition.
- Transposons control their own transposition functions, encoding proteins which act on the element in *cis* or in *trans*.

Bacteriophage Mu infects *E. coli* and consists of a complex, tailed particle containing a linear, double-stranded DNA genome of about 40 kb, with host cell-derived sequences of between 0.5 and 3.0 kbp attached to the right-hand end of the genome (Figure 3.15). Mu is a **temperate bacteriophage** whose replication can proceed through two pathways, one involving integration of the genome into that of the host cell and resulting in **lysogeny**, the other **lytic** replication resulting in the death of the cell (see Chapter 5). Integration of the phage genome into that of the host bacterium occurs at random sites in the cell genome. Integrated phage genomes are known as **prophage** and integration is essential for the establishment of

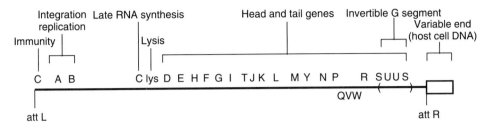

Figure 3.15 Organization of the bacteriophage Mu genome.

lysogeny. At intervals in bacterial cells lysogenic for Mu, the prophage undergoes transposition to a different site in the host genome. The mechanism leading to transposition is different to that responsible for the initial integration of the phage genome (which is conservative, i.e. does not involve replication), and is a complex process, requiring numerous phage-encoded and host cell proteins. Transposition is tightly linked to replication of the phage genome and results in the formation of a 'co-integrate,' i.e. a duplicated copy of the phage genome flanking a target sequence in which insertion has occurred. The original Mu genome remains in the same location where it first integrated and is joined by a second integrated genome at another site. (NB Not all prokaryotic transposons utilize this process. Some, such as TN10, are not replicated during transposition but are excised from the original integration site and integrate elsewhere.) There are two consequences of such a transposition, first that the phage genome is replicated during this process (advantageous for the virus) and second, that the sequences flanked by the two phage genomes (which form repeated sequences) are at risk of secondary rearrangements, including deletions, inversions, duplications, and translocations (possibly but not necessarily deleterious for the host cell).

Ty is representative of a class of genetic elements found in yeast and other eukaryotes which are known as retrotransposons. Unlike bacteriophage Mu, such elements are not true virus genomes, but bear striking similarities to retroviruses. The genomes of most strains of *Saccharomyces cerevisiae* contain 30–35 copies of the Ty element, which is 5.9 kbp long and contains a 340 bp direct repeat at each end (Figure 3.16). Within this repeat sequence is a promoter which results in the transcription of a terminally redundant 5.7 kb mRNA. This contains two genes, *TYA*, which has homology to the *gag* gene of retroviruses, and *TYB*, which is homologous to the *pol* gene. The protein encoded by *TYA* is capable of forming a roughly spherical, 60 nm diameter 'virus-like particle' (VLP). The 5.7 kb RNA transcript can be incorporated into such particles, resulting in the formation of intracellular structures known as Ty-VLPs. These particles are not infectious for yeast cells like true viruses, but if accidentally taken up by a cell, can carry out reverse transcription of their RNA content to form a double-stranded DNA Ty element which can then integrate into the host cell genome (see below).

The most significant difference between retrotransposons such as Ty, *copia* (a similar element found in *Drosophila melanogaster*), and retroviruses proper is the presence of an additional gene, *env*, which encodes an envelope glycoprotein; there

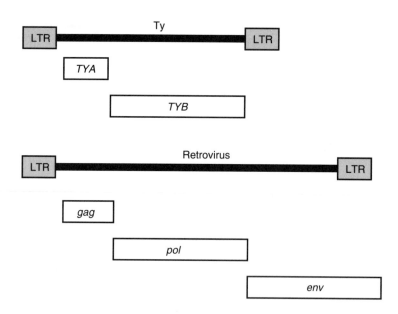

Figure 3.16 The genetic organizations of retrotransposons such as Ty (above) and retrovirus genomes (below) show a number of similarities, including the presence of direct long terminal repeats (LTRs) at both ends.

is also an outer lipid envelope in retroviruses (see Chapter 2). The envelope protein is responsible for receptor binding and has allowed retroviruses to escape the intracellular lifestyle of retrotransposons, to form a true virus particle and propagate themselves widely by infection of other cells (Figure 3.16). As already noted, retroviruses are the only truly diploid viruses, since the particles contain two copies of the genomic RNA. They also contain a cellular tRNA which is used as a primer for DNA synthesis during reverse transcription of the genome (Figure 3.17).

During the process of reverse transcription, the two single-stranded (+)sense RNA molecules which comprise the virus genome are converted into a double-stranded DNA molecule somewhat longer than the RNA templates owing to the duplication of direct repeat sequences at each end – the long terminal repeats (LTRs) (Figure 3.18). Some of the steps in reverse transcription have remained mysterious; for example, the apparent jumps which the polymerase makes from one end of the template strand to the other. In fact, these steps can be explained by the observation that complete conversion of retrovirus RNA into double-stranded DNA only occurs in a partially uncoated core particle and cannot be duplicated accurately *in vitro* with the reagents free in solution. This indicates that the conformation of the two RNAs inside the retrovirus nucleocapsid dictates the course of reverse transcription – the 'jumps' are nothing of the sort, since the ends of the strands are probably held adjacent to one another inside the core.

Reverse transcription has important consequences for retrovirus genetics. In the first place, it is a highly error-prone process, since reverse transcriptase does not

Figure 3.17 Mechanism of reverse transcription of retrovirus RNA genomes, in which two molecules of RNA are converted into a single (terminally redundant) double-stranded DNA provirus.

Integrated provirus

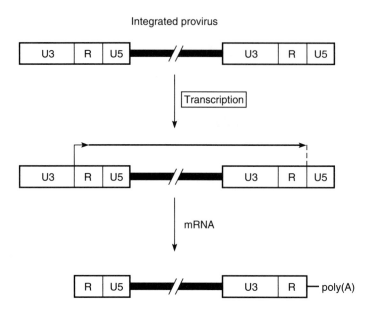

Figure 3.18 Generation of repeated information in retrovirus long terminal repeats. In addition to their role in reverse transcription, these sequences contain important control elements involved in the expression of the virus genome, including a transcriptional promoter in the U3 region and polyadenylation signal in the R region.

carry out the proofreading functions performed by cellular DNA-dependent DNA polymerases. This results in the introduction of many mutations into retrovirus genomes and, consequently, rapid genetic variation (see Spontaneous Mutations, page 52). In addition to this, the process of reverse transcription promotes genetic recombination. Since two RNAs are packaged into each virion and used as the template for reverse transcription, recombination can and does occur between the two strands. Although the mechanism responsible for this is not clear, if one of the RNA strands differs from the other, for example, by the presence of a mutation, and recombination occurs, then the resulting virus will be genetically distinct from either of the parental viruses.

After reverse transcription is complete, the double-stranded DNA migrates into the nucleus, still in association with virus proteins. The mature products of the *pol* gene are, in fact, a complex of polypeptides which include three distinct enzymatic activities: reverse transcriptase and RNAse H, which are involved in reverse transcription, and integrase, which catalyzes integration of virus DNA into the host cell chromatin, following which it is known as the **provirus** (Figure 3.19). Three forms of double-stranded DNA are found in retrovirus-infected cells following

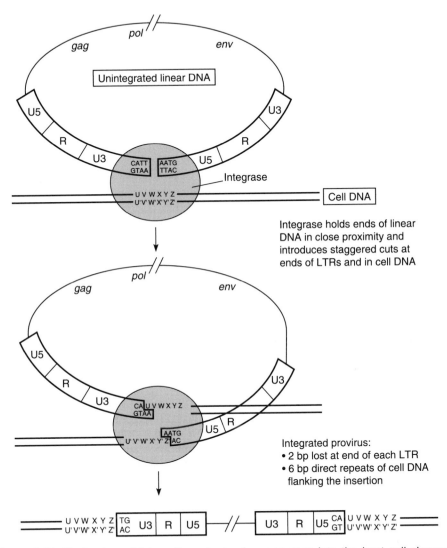

Figure 3.19 Mechanism of integration of retrovirus genomes into the host cell chromatin.

reverse transcription: linear DNA and two circular forms which contain either one or two LTRs. From the structure at the ends of the provirus, it was previously believed that the two-LTR circle was the form used for integration. In recent years, systems which have been developed to study the integration of retrovirus DNA *in vitro* suggest that it is, in fact, the linear form which integrates. This discrepancy can be resolved by a model in which the ends of the two LTRs are held in close proximity by the reverse transcriptase–integrase complex. The net result of integration is that 1–2 bp are lost from the end of each LTR and 4–6 bp of cellular DNA are duplicated on either side of the provirus. It is unclear whether there is any specificity regarding the site of integration into the cell genome. What is obvious is that there is no simple target sequence, but it is possible that there may be (numerous) regions or sites in the eukaryotic genome which are more likely to be integration sites than others.

Following integration, the DNA provirus genome becomes essentially a collection of cellular genes and is at the mercy of the cell for expression. There is no mechanism for the precise excision of integrated proviruses, some of which are known to have been 'fossilized' in primate genomes through millions of years of evolution, although proviruses may sometimes be lost or altered by modifications of the cell genome. The only way out for the virus is transcription, forming what is essentially a full-length mRNA (minus the terminally redundant sequences from the LTRs). This RNA is the vRNA and two copies are packaged into virions (Figure 3.18).

There are, however, two different genome strategies used by viruses which involve reverse transcription. It is at this point that the difference between them becomes obvious. One strategy, as used by retroviruses and described above, culminates in the packaging of RNA into virions as the virus genome. The other, used by hepadnaviruses and caulimoviruses, switches the RNA and DNA phases of replication and results in DNA virus genomes. This is achieved by utilizing reverse transcription not as an early event in replication as retroviruses do, but as a late step during the formation of the virus particle.

Hepatitis B virus (HBV) is the prototype member of the family *Hepadnaviridae*. HBV virions are spherical, lipid-containing particles, 42–47 nm in diameter which contain a partially double-stranded ('gapped') DNA genome, plus an RNA-dependent DNA polymerase (i.e. reverse transcriptase) (Figure 3.20). Hepadnaviruses have the smallest genomes of all known viruses with a (−)sense strand of 3.0–3.3 kb (varies between different hepadnaviruses) and a (+)sense strand of 1.7–2.8 kb (varies between different particles). On infection of cells, three major genome transcripts are produced: 3.4, 2.4, and 2.1 kb mRNAs. All have the same polarity (i.e. are transcribed from the same strand of the virus genome) and the same 3' ends, but have different 5' ends (i.e. initiation sites). These transcripts are heterogeneous in size and it is not completely clear which proteins each transcript encodes, but there are four known genes in the virus:

- C – encodes the core protein.
- P – encodes the polymerase.
- S – encodes the three polypeptides of the surface antigen, pre-S1, pre-S2, and S (which are derived from alternative start sites).
- X – encodes a transactivator of viral transcription (and possibly cellular genes?).

Closed circular DNA is found soon after infection in the nucleus of the cell and is probably the source of the above transcripts. This DNA is produced by repair of the gapped virus genome as follows:

(1) Completion of the (+)sense strand.
(2) Removal of a protein primer from the (−)sense strand and an oligoribonucleotide primer from the (+)sense strand.

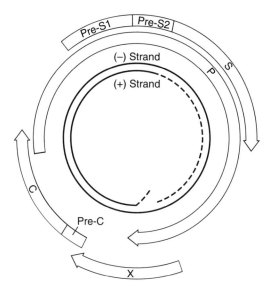

Figure 3.20 Structure, organization, and protein-coding potential of the hepatitis B virus genome.

(3) Elimination of terminal redundancy at the ends of the (−)sense strand.
(4) Ligation of the ends of the two strands.

It is not known how or by which proteins (viral or cellular) these events are carried out. The 3.4 kb RNA transcript, core antigen and polymerase form core particles and it is believed that the polymerase converts the RNA to DNA in these particles as they form in the cytoplasm.

The genome structure and replication of cauliflower mosaic virus (CaMV), the prototype member of the *Caulimovirus* group, is reminiscent of that of hepadna-viruses, although there are differences between them. The CaMV genome consists of a gapped, circular double-stranded DNA molecule of about 8 kbp, one strand of which is known as the α strand and contains a single gap, and a complementary strand which contains two gaps (Figure 3.21). There are eight genes encoded in this genome, although not all eight products have been detected in infected cells. Replication of the CaMV genome is similar to that of HBV. The first stage is the migration of the gapped virus DNA to the nucleus of the infected cell where it is repaired to form a covalently closed circle. This DNA is transcribed to produce two polyadenylated transcripts, one long and one shorter one. In the cytoplasm, the shorter mRNA is translated to produce a protein which forms large **inclusion bodies** in the cytoplasm of infected cells and it is in these sites that the second phase of replication occurs. In these replication complexes, the long mRNA is translated, reverse transcribed, and packaged into virions as they form.

The differences between these viruses and retroviruses are summarized in Table 3.2.

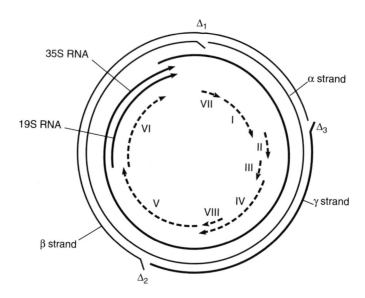

Figure 3.21 Structure, organization and protein-coding potential of the cauliflower mosaic virus genome.

Table 3.2 Comparison between hepadnaviruses, caulimoviruses, and retroviruses

Features	Hepadnaviruses	Caulimoviruses	Retroviruses
Genome	DNA	DNA	RNA
Primer for (−)strand synthesis	Protein	tRNA	tRNA
Terminal repeats	No	No	Yes
Specific integration of virus genome	No	No	Yes

Evolution and Epidemiology

The power of the new tools of molecular biology has given rise to new avenues of investigation in virology. Epidemiology is concerned with the distribution of disease and the derivation of strategies to reduce or prevent it. Virus infections present considerable difficulties for this process. Aside from epidemics where acute symptoms are obvious, the major evidence of virus infection available to the epidemiologist is the presence of anti-viral antibodies in patients. This information frequently provides an incomplete picture, and it is often difficult to assess whether a virus infection occurred recently or at some time in the past. Alternative means of investigation, such as the inoculation of viruses into experimental plants or animals to assess their phenotype, are laborious and impossible to apply to very large populations. Molecular biology provides sensitive, rapid, and sophisticated techniques to detect and analyze the genetic information stored in virus genomes and has resulted in a new area of investigation, molecular epidemiology.

One drawback of molecular genetic analysis is that it assumes a good deal of knowledge about the nature of a virus genome before it can be investigated, i.e. you have to know what you are looking for before you can find it. However, it should be obvious from this chapter that we now possess a great amount of information about the structure and nucleotide sequences of at least representative members of most of the known virus groups. This type of information allows virologists to look in two directions: back to where viruses came from and forward to chart the course of future epidemics and diseases. There is no space here to give details of the methodology used by virologists to achieve these objectives, but readers are referred to Chapter 1 and the references therein. Sensitive detection of nucleic acids by amplification techniques such as the polymerase chain reaction has already and will no doubt continue to have a major impact on this type of epidemiologic investigation.

There are (at least!) three theories which seek to explain the origin of viruses:

- *Regressive evolution* This theory states that viruses are degenerate life-forms which have lost many functions that other organisms possess and have only retained the genetic information essential to their parasitic way of life.
- *Cellular origins* In this theory viruses are thought to be subcellular, functional assemblies of macromolecules which have escaped their origins inside cells.
- *Independent entities*: This says that viruses evolved on a parallel course to cellular organisms from the self-replicating molecules believed to have existed in the primitive prebiotic 'RNA world.'

While each of these theories has its devotees and this subject provokes fierce disagreements, the fact of the matter is that viruses exist, and we are all infected with them. As Max Perutz wrote, 'Who can guess? We were not there when it happened.'

The importance of the origin of viruses is that this issue may have practical implications for virology here and now. Genetic and nucleotide sequence relationships between viruses can reveal the origins not only of individual viruses, but also of whole families and possible 'superfamilies' (Figure 3.22). In a number of groups of viruses previously thought to be unrelated, nucleotide sequencing has revealed that functional regions appear to be grouped together in a similar way. The extent to which there is any sequence similarity between these regions in different viruses varies, although clearly, the active sites of enzymes such as viral replicases are strongly conserved. The emphasis in these groupings is more on functional and organizational similarities. The official classification scheme for viruses does not presently recognize a higher level grouping than the 'family' (see Appendix 1). The one exception to this is the grouping of negative-stranded RNA virus families, the *Mononegvirales*, which embraces the three families of non-segmented negative-stranded viruses, the *Filoviridae*, *Paramyxoviridae*, and *Rhabdoviridae*. Super-families are thus equivalent to the 'orders' of formal biological nomenclature.

Such ideas may allow us to predict the properties and behavior of new viruses, or to develop new drugs based on what is already known about existing viruses. Whether these shared patterns suggest the descent of present-day viruses from a

Positive-stranded viruses

Families:
Picornaviridae, Potyvirus (non-segmented)
Comovirus, Nepovirus (segmented)

Sindbis virus-like

Families:
Togaviridae, Tobravirus, Tobamovirus (non-segmented)
Bromovirus (segmented)

Negative-stranded viruses

Families:
Paramyxoviridae, Rhabdoviridae (non-segmented)
Bunyaviridae, Arenaviridae (segmented)

Figure 3.22 Similarities in the function and organization of the genomes of different virus families allow them to be grouped together as 'superfamilies.'

limited number of primitive ancestors or are evidence of convergent evolution between different virus families is uncertain. Although it is tempting to speculate on events which may have occurred before the origins of 'life' as it is presently recognized, it would be unwise to discount the pressures which might result in viruses with diverse origins assuming common genetic solutions to common problems of storing, replicating, and expressing genetic information. This is particularly true now that we appreciate the plasticity of virus and cellular genomes and the mobility of genetic information from virus to virus, cell to virus, and virus to cell.

Summary

Molecular biology has put much emphasis on the structure and function of the virus genome. At first sight, this tends to emphasize the tremendous diversity of virus genomes. On closer examination, similarities and unifying themes become more apparent. Sequences and structures at the ends of virus genomes are in some ways functionally more significant than the unique coding regions within them. Common patterns of genetic organization seen in virus superfamilies suggest either that many viruses have evolved from common ancestors or that convergent evolution and exchange of genetic information between viruses has resulted in common solutions to common problems.

Further Reading

Adams, R.L.P. (1986). *The Biochemistry of the Nucleic Acids* (10th edition). Chapman and Hall, New York.

Alberts, B., Bray, D., Lewis, J., Raff, M., Roberts, K. and Watson, J.D. (Eds) (1989). *Molecular Biology of the Cell* (2nd edition). Garland, New York.

Fields, B.N. and Knipe, D.M. (Eds) (1990). *Fields' Virology* (2nd edition). Raven Press, New York, chaps 6–9, pp. 95–190.

Bibliography

Baltimore, D. (1985). Retroviruses and retrotransposons: the role of reverse transcription in shaping the eukaryotic genome. *Cell* **40**: 481–2.

Belfort, M. (1989). Bacteriophage introns: parasites within parasites? *Trends in Genetics* **5**: 209–13.

Berget, S.M. *et al.* (1978). Spliced segments at the 5′ termini of adenovirus-2 late mRNA: a role for heterogeneous nuclear RNA in mammalian cells. *Cold Spring Harbor Symp. Quant. Biol.* **42**: 523–9.

Clapham, P. *et al.* (1983). Productive infection and cell free transmission of HTLV in a non-lymphoid cell line. *Science* **222**: 1125–7.

Fuller, F. *et al.* (1980). Sindbis virus: gene order of translation *in vivo*. *Virology* **107**: 441–51.

Holland, J.J. (Ed.) (1991). Genetic diversity of RNA viruses. *Curr. Topics Microbiol. Immunol.* **176**.

Kates, J. (1970). Transcription of the vaccinia virus genome and the occurrence of polyadenylic acid sequences in the messenger RNA. *Cold Spring Harbor Symp. Quant. Biol.* **35**: 743–52.

Sanger, F. *et al.* (1977). Nucleotide sequence of bacteriophage ϕX174. *Nature* **265**: 687–95.

Symonds, N., Toussaint, A., Van de Putte, P. and Howe, M. (Eds) (1987). *Phage Mu*. Cold Spring Harbor Laboratory.

Taber *et al.* (1971). Effect of pactamycin on synthesis of poliovirus proteins: a method for genetic mapping. *J. Virol.* **8**: 395–401.

Chapter 4

Replication

Overview of Virus Replication

The way in which viruses are classified has altered as our perception of them has changed:

(1) *By disease*: Many early civilizations, such as those of ancient Egypt and Greece, were well aware of the pathogenic effects of many different viruses. From these times, we have several surprisingly accurate decriptions of diseases of humans, animals, and crops, although of course the nature of the agents responsible for these calamities was not realized at the time. Accurate though these decriptions were, a major problem with such a system of classification is that many diverse viruses cause similar symptoms; for example, respiratory infections with fever may be caused by many different viruses.

(2) *By morphology*: As increasing numbers of viruses were isolated and techniques of analysis were improved, it became possible during the 1930s–1950s to classify viruses based on the structure of virus particles. Although this is an improvement on the above scheme, there are still problems in distinguishing between viruses that are morphologically similar but cause disparate clinical symptoms (e.g. the various picornaviruses). During this era, serology became an important aid in virus classification, and particle morphology continues to be an important strand in virus classification.

(3) *Functional classification*: In recent years, more emphasis has been placed on the replication strategy of the virus. This is particularly true with respect to the composition and structure of the virus genome and the constraints which this imposes on replication. This approach has accelerated since the development of molecular biology due to the central importance of the virus genome in this new technology. Molecular analysis of virus genomes permits rapid and unequivocal identification of individual virus strains, but is also predictive of the properties of a previously unknown or novel virus with a familiar genome structure. (The classification of known infectious agents is described in Appendix 1.)

In a teleological sense (i.e. crediting an inanimate organism such as a virus with a conscious purpose), the sole 'aim' of a virus is to replicate its genetic information. The nature of the virus genome is therefore preeminent in determining what steps are necessary to achieve this aim. In reality, there is a surprising amount of variation that can occur in these processes, even for viruses with seemingly similar genome structures. The reason for this lies in compartmentalization, both of eukaryotic cells into nuclear and cytoplasmic compartments, and of genetic information and biochemical capacity between the virus genome and that of the host cell.

The type of cell infected by the virus has a profound effect on the process of replication. For viruses of prokaryotes, replication to some extent reflects the relative simplicity of their host cells. For viruses with eukaryotic hosts, matters are frequently more complex. There are many examples of animal viruses undergoing different replicative cycles in different cell types. However, the coding capacity of the genome forces all viruses to choose a strategy for replication. This might be one involving heavy reliance on the host cell, in which case the virus genome can be very compact and need only encode the essential information for a few proteins, e.g. parvoviruses. Alternatively, large and complex virus genomes, such as those of poxviruses, encode most of the information necessary for replication, and the virus is only reliant on the cell for the provision of energy and the apparatus for macromolecular synthesis, such as ribosomes (see Chapter 1). Viruses with an RNA lifestyle, i.e an RNA genome plus messenger RNAs, have no apparent need to enter the nucleus, although during the course of replication, many do. 'DNA viruses,' as might be expected, mostly replicate in the nucleus where host cell DNA is replicated and where the biochemical apparatus necessary for this process is located. However, some viruses with DNA genomes (e.g. poxviruses) have evolved to contain sufficient biochemical capacity to be able to replicate in the cytoplasm, with minimal requirement for host cell functions. Most of this chapter will examine the process of virus replication and will look at some of the variations on the basic theme.

Investigation of Virus Replication

As already stated, bacteriophages have long been used by virologists as models to understand the biology of viruses. This is particularly true of virus replication. During the first half of this century, two particularly significant experiments which illustrated the fundamental nature of viruses were performed on bacteriophages. The first of these was accomplished by Ellis and Delbrück in 1939 and is usually referred to as the 'single-burst' experiment or 'one-step growth curve' (Figure 4.1). This was the first experiment to show the three esssential phases of virus replication:

- Initiation of infection
- Replication and expression of the virus genome
- Release of mature virions from the infected cell.

In this experiment, bacteriophages were added to a culture of rapidly growing bacteria and after a period of a few minutes, the culture was diluted, effectively preventing further interaction between the phage particles and the cells. This simple step is actually the key to the entire experiment, since it effectively synchronizes the infection of the cells and allows the subsequent phases of replication in a population of individual cells and virus particles to be viewed as if it were a single

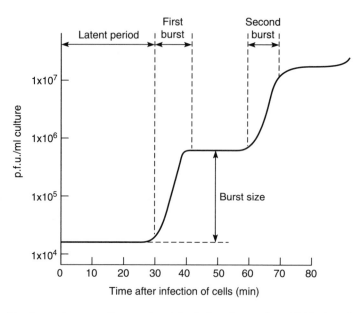

Figure 4.1 The 'one-step growth curve' or 'single-burst experiment.' First performed by Ellis and Delbrück in 1939, this classic experiment illustrates the true nature of virus replication. Details of the experiment are described in the text.

interaction (in much the same way that molecular cloning of nucleic acids allows analysis of populations of nucleic acid molecules as single species). Repeated samples of the culture were taken at short intervals and analyzed for bacterial cells by plating onto agar plates and for phage particles by plating onto lawns of bacteria. As can be seen in Figure 4.1, there is a stepwise increase in the concentration of phage particles with time, each increase in phage concentration representing one replicative cycle of the virus. However, the data from this experiment can also by analyzed in a different way, by plotting the number of **plaque-forming units** (p.f.u.) per bacterial cell against time (Figure 4.2). In this type of assay, a plaque forming unit can be either a single extracellular virus particle or an infected bacterial cell. These two can be distinguished by disruption of the bacteria with chloroform before plating, which releases any intracellular phage particles, giving the total virus count, i.e. intracellular plus extracellular particles.

Several additional features of virus replication are visible from the graph in Figure 4.2. Immediately after dilution of the culture, there is a phase of 10–15 min when no phage particles are detectable; this is known as the **eclipse period**. This represents a time when virus particles have broken down after penetrating cells, releasing their genomes within the host cell as a prerequisite to replication. At this stage, they are no longer infectious and therefore cannot be detected by the plaque assay. The **latent period** is the time before the first new extracellular virus particles appear and is of the order of 20–25 min for most bacteriophages. About 40 min after the cells were infected, the curves for the total number of virus

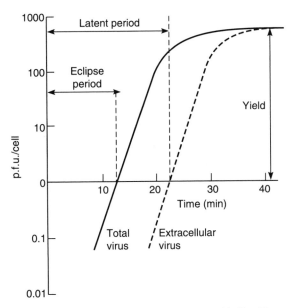

Figure 4.2 Analysis of data from a 'single-burst' experiment. Unlike Figure 4.1, which shows the total number of **plaque-forming units** (p.f.u.) produced, here the data are plotted as p.f.u./bacterial cell, which reflects the events occurring in a 'typical' infected cell in the population. The phases of replication named on the graph are defined in the text.

particles and for extracellular virus merge, since the infected cells have lyzed and released any intracellular phage particles by this time. The yield (i.e. number) of particles produced per infected cell can be calculated from the overall rise in phage titer.

Following the development of plaque assays for animal viruses in the 1950s, 'single burst' experiments have now been performed for many viruses of eukaryotes with similar results (Figure 4.3). The major difference between these viruses and bacteriophages is the much longer time interval required for replication, which is measured in terms of hours and, in some cases, days, rather than minutes after infection. This difference reflects the much slower growth rate of eukaryotic cells, and in part, the complexity of viral replication in compartmentalized cells. Biochemical analysis of virus replication in eukaryotic cells has also been used to analyze the levels of viral and cellular protein and nucleic acid synthesis and to examine the intracellular events occurring during synchronous infections (Figure 4.4). The application of various metabolic inhibitors has also proved to be a valuable tool in such experiments. Examples of the use of such drugs will be given during the detailed consideration of the stages of virus replication later in this chapter.

The second key experiment on virus replication using bacteriophages was performed by Hershey and Chase in 1952. Bacteriophage T2 was propagated in *E. coli* cells which had been 'labeled' with one of two radioisotopes, either ^{35}S, which is incorporated into sulfur-containing amino acids in proteins, or ^{32}P, which is incorporated into nucleic acids (which do not contain any sulfur) (Figure 4.5).

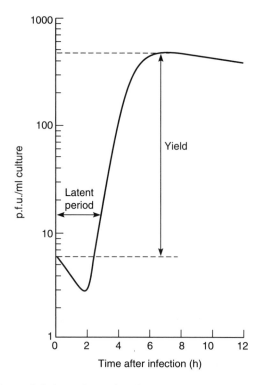

Figure 4.3 Replication of lytic eukaryotic viruses occurs in a way similar to that of bacteriophages. This figure shows a 'single-burst' type of experiment for a picornavirus (e.g. poliovirus). This type of data can only be produced from synchronous infections where a high **multiplicity of infection** is used.

Particles labeled in each of these ways were used to infect bacteria. After a short time to allow attachment to the cells, the mixture was homogenized briefly in a blender which did not destroy the bacterial cells but was sufficiently vigorous to knock the phage coats off the outside of the cells. Analysis of the radioactive content in the cell pellets and culture supernatant (containing the empty phage coats) showed that most of the radioactivity in the ^{35}S-labeled particles remained in the supernatent, while in the ^{32}P-labeled particles, most of the radiolabel had entered the cells. This experiment indicated that it was the DNA genome of the bacteriophage that entered the cells and initiated the infection.

Although this might seem obvious now, at the time this experiment settled a great controversy over whether a structurally simple polymer such as a nucleic acid, which was known to contain only four monomers, was complex enough to carry genetic information. (At the time, it was generally believed that proteins, which consist of a much more complex mixture of more than 20 different amino acids, were the carriers of the genes and that DNA was probably a structural

Figure 4.4 Biochemistry of virus infection. This graph shows the rate of cellular and viral DNA synthesis (based on the incorporation of radio-labeled nucleotides into high molecular weight material) in uninfected and in herpesvirus-infected cells.

component of cells and viruses.) Together, these two experiments illustrate the process of virus replication. Virus particles enter susceptible cells and release their genomic nucleic acids. These are replicated and packaged into virus particles consisting of newly synthesized virus proteins, which are then released from the cell.

The Replication Cycle

Virus replication can be divided into eight stages, as shown in Figure 4.6. It should be emphasized that these are purely arbitrary divisions, used here for convenience in explaining the replication cycles of a non-existent 'typical' virus. For the sake of clarity, the examples given in this section are all taken from the replication of animal viruses. Viruses which infect invertebrates (including bacteria) and plants (except cauliflower mosaic virus) have been deliberately left out. Readers who are particularly interested in these latter viruses are advised to consult more specialist sources for the details of their replication. Nevertheless, regardless of their hosts, *all* viruses must undergo each of these stages in some form in order to successfully complete their replication cycles. Not all of the steps described here are detectable as distinct stages for all viruses; often they are blurred together and appear to occur almost simultaneously. Some of the individual stages have been studied in

Figure 4.5 The Hershey–Chase experiment, first performed in 1952, demonstrated that virus genetic information was encoded by nucleic acids and not proteins. Details of the experiment are described in the text.

great detail and a tremendous amount of information is known about them. Other stages are more mysterious, in some cases, virtually a black box.

Attachment

Since the stages of viral replication used here are purely arbitrary, and because complete replication necessarily involves a cycle, it is possible to begin discussion of virus replication at any point. Arguably, it is most logical to consider the first interaction of a virus with a new host cell as the starting point of the cycle. Technically, virus attachment consists of specific binding of a **virus-attachment protein** (or 'antireceptor') to a cellular **receptor** molecule. Many examples of virus receptors are now known (Table 4.1). The target receptor molecules on cell surfaces may be proteins (usually glycoproteins), or the carbohydrate residues present on glycoproteins or glycolipids. The former are usually specific receptors, in that a virus may use a particular protein as a receptor. Carbohydrate groups are usually

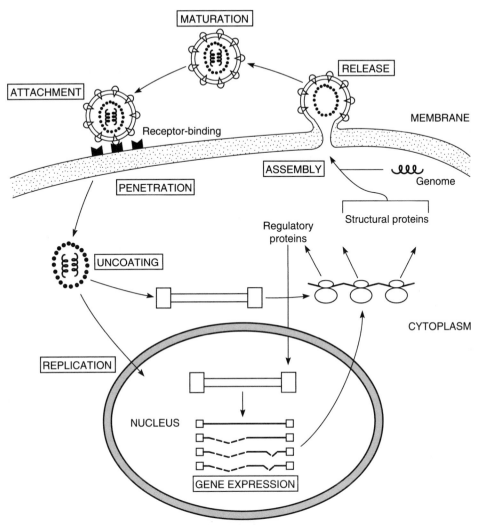

Figure 4.6 Generalized scheme for virus replication.

less specific receptors, since the same configuration of side-chains may occur on many different glycosylated membrane-bound molecules. Some complex viruses (e.g. poxviruses, herpesviruses) may use more than one receptor and therefore have alternative routes of uptake into cells.

Two of the best understood examples of virus–receptor interactions are currently from the picornavirus family and that of influenzavirus. Picornaviruses are unusual in that the virus–receptor interaction has been intensively studied from both angles, i.e. structural features of the virus responsible for receptor binding and those of the receptor molecule itself. The human rhinovirus receptor molecule, ICAM-1 (intracellular adhesion molecule 1), is a cellular adhesion molecule whose normal function is to bind cells to adjacent substrates. Structurally, ICAM-1 is similar to an immunoglobulin molecule, with constant (C) and variable (V) domains homologous

Table 4.1 Known virus receptors and their respective virus-attachment proteins

Virus	Receptor	Virus-attachment protein
RNA viruses		
Picornaviruses		
Poliovirus	100 kDa member of immunoglobulin-like 'superfamily'	VP1
Human rhinovirus	Intracellular adhesion molecule 1 (ICAM-1)	VP1, 2 and 3 (form 'canyon' on surface of virus capsid)
Reoviruses		
Reovirus 3	β-Adrenergic receptor/ sialoglycoproteins	C-Terminal portion of σ-1 protein (hemagglutinin)
Togaviruses		
Semliki forest virus	Class I HLA/H-2 MHC (?)	E2 envelope glycoprotein
Orthomyxoviruses		
Influenzavirus	Sialyloligosaccharides	Hemagglutinin (HA) protein
Paramyxoviruses		
Sendai virus	Sialyloligosaccharides	Hemagglutinin-neuraminidase (HN) protein
Rhabdoviruses		
Rabies virus	Acetylcholine receptor/ Sialylated gangliosides	Virus G protein
Retroviruses		
HIV-1 and HIV-2	CD4 molecule (+ accessory proteins)	gp120 envelope glycoprotein
Visna virus	MHC class II (?)	Envelope glycoprotein
DNA viruses		
Adenoviruses		
Human adenovirus	Class I HLA MHC molecule	Fibre protein
Herpesviruses		
Cytomegalovirus	Class I HLA MHC molecule + β$_2$-microglobulin	α-Domain of H301 protein (Virus-encoded MHC class I-like molecule)
Epstein–Barr virus	C3d receptor 'CR2' (=CD21)	gp 350
Poxviruses		
Vaccinia virus	Epidermal growth factor (EGF) receptor	VGF protein
Hepadnaviruses		
Hepatitis B virus	Hepatocyte receptor for polymeric serum albumin/ polymeric IgA; Sialoglycoprotein	Pre-S portion of envelope protein

icosahedral capsid, which is formed by the flanking monomers, VP1, VP2, and VP3 (Figure 4.8). There is biochemical evidence from a class of inhibitory drugs which block attachment of rhinovirus particles to cells that the interaction between ICAM-1 and the virus particle occurs on the floor of this shielded canyon (Figure 4.9). Unlike other areas of the virus surface, the amino acid residues forming the internal surfaces of the canyon are relatively invariant. They are protected from antigenic pressure because the antibody molecules are too large to fit into the cleft. This is important because radical changes in this region, although allowing the virus to escape an immune response, would disrupt receptor binding. In polioviruses, instead of a canyon, there is a trough which runs around each five-fold vertex of the capsid. The highly variant regions of the capsid to which antibodies bind are located on the 'peaks' on either side of this trough, which once again is too narow to allow antibody binding to the residues at the base of the trough required for the interaction of the capsid with the receptor molecule.

The influenza virus hemagglutinin protein forms one of the two types of glycoprotein spikes visible on the surface of the particles (see Chapter 2), the other type being formed by the neuraminidase protein. Each hemagglutinin spike is composed of a trimer of three molecules, while the neuraminidase spike consists of a tetramer (Figure 4.10). The hemagglutinin spikes are responsible for binding the influenza virus receptor, which is sialic acid (N-acetyl neuraminic acid), a sugar group commonly found on a variety of glycosylated molecules. As a result, there is little cell-type specificity imposed by this receptor interaction and therefore influenza viruses bind to a wide variety of different cell types (e.g. causing

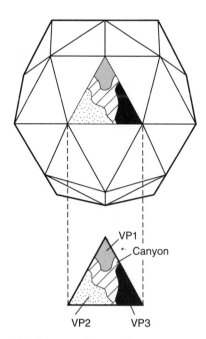

Figure 4.8 Rhinovirus particles have a deep surface cleft known as the 'canyon' between the three monomers (VP1, 2 and 3) making up each face of the particle.

of antibodies and is regarded as a member of the immunoglobuli

 _ily of proteins (Figure 4.7). Similarly, the poliovirus receptor is an integra

_e protein which is also a member of this family, with one variable an

_tant domains. Unlike ICAM-1, the normal function of the poliovirus recepto

is not known, but a murine homolog has recently been identified which is structurally

similar to the human molecule but does not function as a receptor for poliovirus.

However, the existence of two conserved molecules offers an opportunity for

studying the molecular interaction of poliovirus with its receptor.

Since the structure of a number of picornavirus capsids is known at a resolution

of a few ångströms (Chapter 2), it has been possible in recent years to determine

the features of the virus responsible for receptor binding. In rhinoviruses, there is

a deep cleft known as the 'canyon' in the surface of each triangular face of the

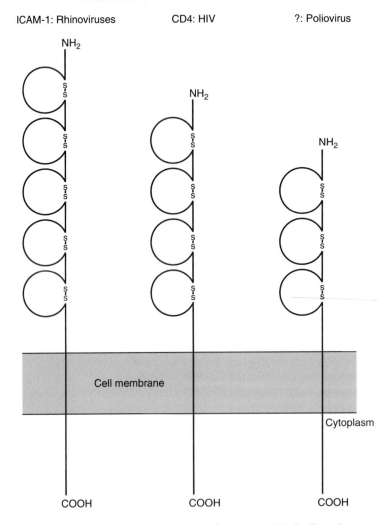

Figure 4.7 Schematic diagram of the structure of immunoglobulin-like virus receptors. A number of receptors belonging to this superfamily of proteins have now been identified.

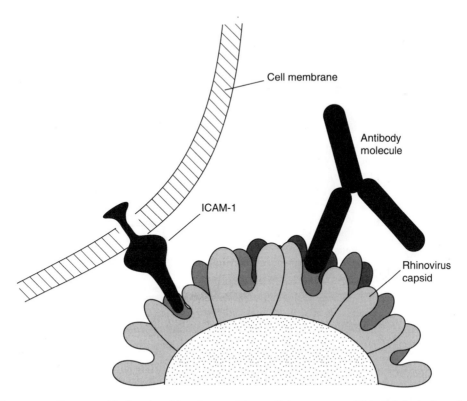

Figure 4.9 Receptor-binding by rhinoviruses. The cellular receptor (ICAM-1) is believed to interact with residues at the bottom of the 'canyon' on the surface of the virus particle (Figure 4.8). Antibody molecules are too large to penetrate the canyon, allowing the conserved receptor-binding site to avoid becoming a target for a neutralizing response.

hemagglutination of red blood cells) in addition to the cells in which productive infection occurs.

The neuraminidase molecule of influenza virus and paramyxoviruses illustrates another feature of this stage of virus replication. Attachment to cellular receptors is in most cases a reversible process – if penetration of the cells does not ensue, the virus can elute from the cell surface. Some viruses have specific mechanisms for detachment and the neuraminidase protein is one of these. Neuraminidase is an esterase which cleaves sialic acid from sugar side-chains. This is particularly important for influenza. Because the receptor molecule is so widely distributed, the virus tends to bind inappropriately to a variety of cells and even cell debris. However, elution from the cell surface after receptor binding has occurred often leads to changes in the virus (e.g. loss or structural alteration of virus-attachment protein) which decrease or eliminate the possibility of subsequent attachment to other cells. Thus in the case of influenza, cleavage of sialic acid residues by neuraminidase leaves these groups bound to the active site of the hemagglutinin, preventing that particular molecule from binding to another receptor.

In many if not most cases, the expression (or absence) of receptors on the

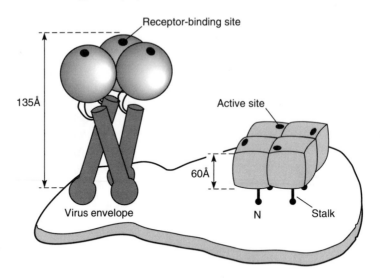

Figure 4.10 Influenza virus glycoprotein spikes.

surface of cells largely determines the **tropism** of a virus, i.e. the type of host cell in which it is able to replicate. In a few cases, intracellular blocks at later stages of replication are responsible for determining the range of cell types in which a virus can carry out a productive infection, but this is not common. Therefore, this initial stage of replication and the very first interaction between the virus and the host cell has a major influence on viral pathogenesis and in determining the course of a virus infection.

An example of the importance of receptor-determined tropism is seen in the replication of Epstein–Barr virus (EBV). EBV only infects human cells, but within the body can infect two cell types: epithelial cells, in which a **productive infection** occurs, and B-lymphocytes, which are usually non-productive. The infection of B-lymphocytes is known to occur via the CR2 molecule on the surface of the cell. However, although they are infected by an unknown route *in vivo*, most epithelial cell lines lack CR2 and are therefore not infectable *in vitro*. This block can be overcome by **transfection** of epithelial cells with recombinant DNA constructions that express CR2; this renders the cell permissive for EBV infection.

A similar observation has been made with HIV. The primary receptor for this virus is another lymphocyte cell surface molecule, CD4. Transfection of human cells which do not normally express CD4 (such as epithelial cells) with recombinant CD4-expression constructs renders them permissive for HIV infection. However, in this case, transfection of rodent cells with human CD4-expression vectors does not permit productive HIV infection. If HIV proviral DNA is inserted into these rodent cells by transfection, virus is produced, indicating that there is no intracellular block to infection. Therefore, it has been proposed that there are one or more (human) proteins in addition to CD4 which are required to form a functional HIV receptor. Furthermore, there is evidence that in at least some cells types, HIV infection is not blocked by competing soluble CD4, indicating that in these cells, a different receptor molecule (or at least, a different route of virus uptake, see below) may be being used.

In some cases, specific receptor binding can be side-stepped by non-specific or inappropriate interactions between virus particles and cells. It is possible that virus particles can be 'accidentally' taken up by cells via processes such as pinocytosis or phagocytosis (see below). However, in the absence of some form of physical interaction which holds the virus particle in close association with the cell surface, the frequency with which these accidental events happen is very low. One such case of an alternative interaction to that involving a specific virus receptor is that of antibody-coated virus particles binding to Fc receptor molecules on the surface of monocytes and other blood cells, which can result in virus uptake. This phenomenon has been shown to occur in a number of cases, where antibody-dependent enhancement of virus uptake results in unexpected findings; for example, the presence of anti-viral antibodies can occasionally result in increased virus uptake by cells and increased pathogenicity, rather than virus neutralization as would normally be expected. It has been suggested that this mechanism may also be important in the uptake of HIV by macrophages and monocytes and that this might be a factor in the pathogenesis of AIDS, although this is not yet certain.

Penetration

Penetration of the target cell normally occurs a very short time after attachment of the virus to its receptor in the cell membrane. Unlike attachment, cell penetration is generally an energy-dependent process, i.e. the cell must be metabolically active for this to occur. Three main mechanisms are involved:

(1) *Translocation* of the entire virus particle across the cytoplasmic membrane of the cell (Figure 4.11). This process is relatively rare among viruses and is poorly understood. It must be mediated by proteins in the virus capsid and specific membrane receptors.

(2) *Endocytosis* of the virus into intracellular vacuoles (Figure 4.12). This is probably the most common mechanism of virus entry into cells. It does not require any specific virus proteins (other than those already utilized for receptor binding) but relies on the normal formation and internalization of coated pits at the cell membrane. Receptor-mediated endocytosis is an efficient process for taking up and concentrating extracellular macromolecules.

(3) *Fusion* of the virus envelope (only applicable to enveloped viruses) with the cell membrane, either directly at the cell surface or following endocytosis in a cytoplasmic vesicle (Figure 4.13). Fusion require the presence of a specific **fusion protein** in the virus envelope, e.g. influenza hemagglutinin or retrovirus TM glycoproteins. These proteins promote the joining of the cellular and virus membranes which results in the nucleocapsid being deposited directly in the cytoplasm. There are two types of virus-driven membrane fusion: one is pH-dependent and the other is pH-independent (see below).

The process of endocytosis is almost universal in animal cells and requires

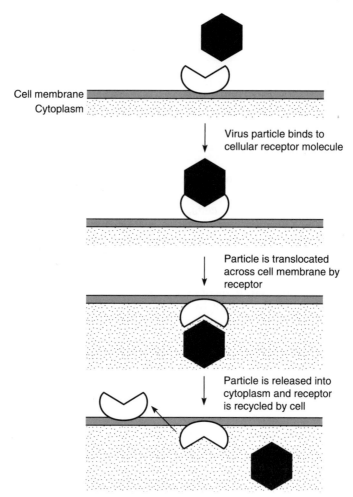

Cell membrane
Cytoplasm

Virus particle binds to
cellular receptor molecule

Particle is translocated
across cell membrane by
receptor

Particle is released into
cytoplasm and receptor
is recycled by cell

Figure 4.11 Translocation of entire virus particles across the cell membrane by cell surface receptors.

further consideration (Figure 4.12). The formation of coated pits results in the engulfment of a membrane-bounded vesicle by the cytoplasm of the cell. The lifetime of these initial coated vesicles is very short. Within seconds, most fuse with endosomes, releasing their contents into these larger vesicles. At this point, any virus contained within these structures is still cut off from the cytoplasm by a lipid bilayer and therefore has not yet really entered the cell. Moreover, as endosomes fuse with episomes and eventually lysosomes, the environment inside these vessels becomes progressively more hostile as they are acidified and the pH falls, while the concentration of degradative enzymes rises. This means that the virus must leave the vesicle and enter the cytoplasm before it is degraded. There are a number of mechanisms by which this occurs, including membrane fusion and rescue by transcytosis. The release of virus particles from endosomes and their

ENDOCYTOSIS

EXOCYTOSIS

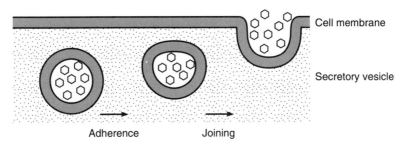

Figure 4.12 Endocytosis and exocytosis of virus particles.

passage into the cytoplasm is intimately connected with (and often impossible to separate from) the process of uncoating (see below).

Uncoating

Uncoating is a general term for the events which occur after penetration, in which the virus capsid is completely or partially removed and the virus genome exposed, usually in the form of a nucleoprotein complex. Unfortunately, uncoating is one of the stages of virus replication that has been least studied and is relatively poorly understood.

In one sense, the removal of a virus envelope which occurs during membrane fusion is part of the uncoating process. Fusion between virus envelopes and endosomal membranes is driven by viral fusion proteins, but these are usually activated by the uncloaking of a previously hidden fusion domain due to conformational changes in the protein induced by the low pH inside the vesicle. The initial events in uncoating may occur inside endosomes, being triggered by the change in pH as the endosome is acidified, or directly in the cytoplasm. Ionophores,

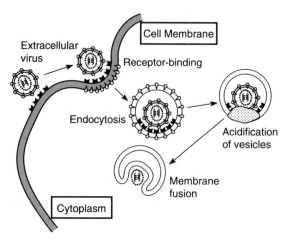

Figure 4.13 Virus-induced membrane fusion. This process is dependent on the presence of a specific **fusion protein** on the surface of the virus which, under particular circumstances (e.g. acidification of the virus-containing vesicle), becomes activated, inducing fusion of the vesicle membrane and the virus envelope.

such as monesin or nigericin, in combination with cations, such as chloroquine and ammonium chloride, can be used to block the acidification of these vesicles and to determine whether events are occurring following the acidification of endosomes (e.g. pH-dependent membrane fusion) or directly at the cell surface or in the cytoplasm (e.g. pH-independent membrane fusion). In picornaviruses, penetration of the cytoplasm by passage of virus from endosomes into the cytoplasm is tightly linked to uncoating. The acidic environment of the endosome causes a conformational change in the capsid which reveals hydrophobic domains not present on the surface of mature virus particles. The interaction of these hydrophobic patches with the endosomal membrane is believed to form pores through which the genome passes into the cytoplasm.

The product of uncoating depends on the structure of the virus nucleocapsid. In some cases, this might be relatively simple (e.g. picornaviruses have a small basic protein of approximately 23 amino acids (VPg) covalently attached to the 5′ end of the vRNA genome), or may be highly complex (e.g. retrovirus cores are highly ordered nucleoprotein complexes which contain, in addition to the diploid RNA genome, the reverse transcriptase enzyme responsible for converting the virus RNA genome into the DNA provirus). The structure and chemistry of the nucleocapsid determines the subsequent steps in replication. As discussed in Chapter 3, reverse transcription can only occur inside an ordered retrovirus core particle and cannot proceed with the components of the reaction free in solution. Herpesvirus, adenovirus, and papovavirus capsids undergo structural changes following penetration, but overall remain largely intact. These capsids contain sequences which are responsible for attachment to the cytoskeleton and this interaction allows the transport of the entire capsid to the nucleus. It is at the nuclear pores that uncoating occurs and the nucleocapsid passes into the nucleus. In reoviruses and poxviruses, complete uncoating does not occur, and many of the reactions of genome replication are catalyzed by virus-encoded enzymes inside cytoplasmic particles which still resemble the mature virions.

Genome replication and gene expression

The replication strategy of any virus depends on the nature of its genetic material, and in this respect, viruses can be divided into seven groups. Such a scheme was first proposed by David Baltimore in 1971. Originally, this classification included only six groups, but it has since been extended to include the scheme of genome replication used by the hepadnaviruses and caulimoviruses. For viruses with RNA genomes in particular, genome replication and the expression of genetic information are inextricably linked. Therefore, both of these criteria are taken into account in the scheme below. The control of gene expression determines the overall course of a virus infection (acute, chronic, persistent, or latent) and such is the emphasis placed on gene expression by molecular biologists that this subject is discussed in detail in Chapter 5. A schematic overview of the major events during replication of the different virus genomes is shown in Figure 4.14.

Class I: Double-stranded DNA (*Papovaviridae; Adenoviridae; Herpesviridae; Poxviridae*) (Figure 4.14a). This class can be subdivided into two further groups:

  (a) Replication is exclusively nuclear (*Papovaviridae; Adenoviridae; Herpesviridae*). The replication of these viruses is relatively dependent on cellular factors.

  (b) Replication occurs in cytoplasm (*Poxviridae*). These viruses have evolved (or acquired) all the necessary factors for transcription and replication of their genomes and are therefore largely independent of the cellular machinery.

Class II: Single-stranded (+)sense DNA (*Parvoviridae*) (Figure 4.14b). Replication occurs in the nucleus, involving the formation of a (−)sense strand, which serves as a template for (+)strand synthesis.

Class III: Double-stranded RNA (*Reoviridae*) (Figure 4.14c). These viruses have segmented genomes. Each segment is transcribed separately to produce individual monocistronic mRNAs.

Class IV: Single-stranded (+)sense RNA (*Picornaviridae; Caliciviridae; Togaviridae; Flaviviridae; Coronaviridae*). These can be subdivided into two groups:

  (a) Viruses with polycistronic mRNA (*Picornaviridae; Caliciviridae; Flaviviridae*) (Figure 4.14d). As with all the viruses in this class, the genome RNA forms the mRNA which is translated after infection, resulting in the synthesis of a polyprotein product, which is subsequently cleaved to form the mature proteins.

  (b) Viruses with complex transcription (*Togaviridae; Coronaviridae*) (Figure 4.14e). Two rounds of translation are necessary to produce the genomic RNA.

Class V: Single-stranded (−)sense RNA. (*Orthomyxoviridae; Paramyxoviridae; Rhabdoviridae; Filoviridae; Bunyaviridae*) (Figure 14.4f). As discussed in Chapters 3 and 5, the genomes of these viruses can be divided into two types:

  (a) Segmented (*Orthomyxoviridae*). The first step in replication is transcription

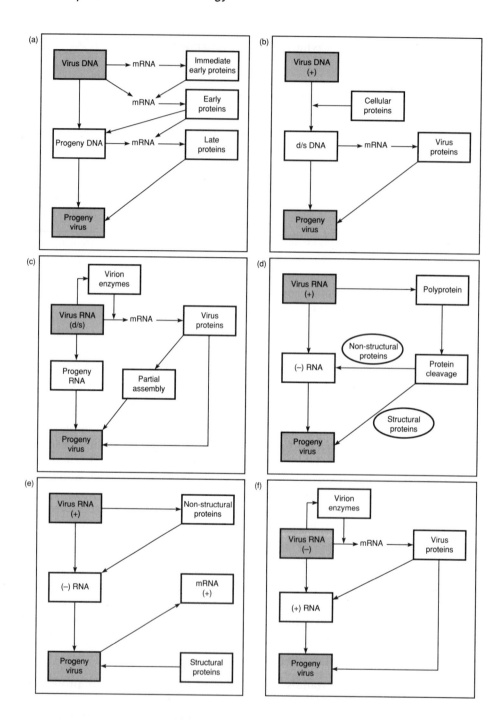

Figure 4.14 Schematic representation of the pattern of replication of virus genomes. Details of the events which occur for genomes of each type are given in the text.

(g)

(h)

Figure 4.14 *continued.*

of the (−)sense RNA genome by the virion RNA-dependent RNA polymerase to produce monocistronic mRNAs, which also serve as the template for subsequent genome replication.

(b) *Non-segmented.* Replication occurs as above, with monocistronic mRNAs for each of the virus genes produced by the virus transcriptase from the full-lengh virus genome (see Chapter 5).

Class VI: Single-stranded (+)sense RNA with DNA intermediate (*Retroviridae*) (Figure 4.14g). The virus genome is (+)sense RNA but unique among viruses in two respects, first that it is diploid, and second, that it does not serve as mRNA, but as a template for reverse transcription into DNA (see Chapter 3).

Class VII: Double-stranded DNA with RNA intermediate (*Hepadnaviridae; Caulimovirus*) (Figure 4.14h). This group of viruses also relies on reverse transcription, but unlike the retroviruses, this occurs inside the virus particle during maturation. On infection of a new cell, the first event to occur is repair of the gapped genome, followed by transcription (see Chapter 3).

Assembly

This process involves the collection of all the components necessary for the formation of the mature virion at a particular site in the cell. During assembly, the basic structure of the virus particle is formed. The site of assembly depends on the site of replication within the cell and on the mechanism by which the virus is eventually released from the cell and varies for different viruses; for example in picornaviruses, poxviruses, and reoviruses it occurs in the cytoplasm; in adenoviruses, papovaviruses, and parvoviruses it occurs in the nucleus; and in retroviruses it occurs on the inner surface of the cell membrane.

As with the early stages of replication, it is not always possible to identify the assembly, maturation, and release of virus particles as distinct and separate phases. In all cases, however, the site of assembly has a profound influence on all these processes. In the vast majority of cases, cellular membranes are used to anchor virus proteins, and this initiates the process of assembly. In spite of considerable study, the control of virus assembly is generally not well understood. In general, it is thought that rising intracellular levels of virus proteins and genome molecules reach a critical concentration and this triggers the process. Many viruses achieve high levels of newly synthesized structural components by concentrating these into subcellular compartments, visible in light microscopes, which are known as **inclusion bodies**. These are a common feature of the late stages of infection of cells by many different viruses, and the size and location of inclusion bodies in infected cells is often highly characteristic of particular viruses; for example, rabies virus infection results in large perinuclear 'Negri bodies,' first observed by Adelchi Negri in 1903. Alternatively, local concentrations of viral structural components can be boosted by lateral interactions between membrane-associated proteins. This mechanism is particularly important in eveloped viruses which are released from the cell by budding (see below).

As discussed in Chapter 2, the formation of virus particles may be a relatively simple process which is driven only by interactions between the subunits of the capsid and controlled by the rules of symmetry. In other cases, assembly is a highly complex, multi-step process involving not only virus structural proteins, but in addition virus-encoded and cellular scaffolding proteins which act as templates to guide the assembly of virions. The encapsidation of the virus genome may occur either early in the assembly of the particle (e.g. many viruses with helical symmetry are nucleated on the genome), or at a late stage, when the genome is stuffed into an almost completed protein shell.

Maturation

This is the stage of the life-cycle at which the virus becomes infectious. Maturation usually involves structural changes in the virus particle which mày result from specific cleavages of capsid proteins to form the mature products or conformational changes in proteins during assembly. Such events frequently lead to substantial structural changes in the capsid which may be detectable by criteria such as differences in the antigenicity of incomplete and mature virus particles, which in some cases (e.g. picornaviruses) alters radically. Alternatively, internal structural alterations, for example, the condensation of nucleoproteins with the virus genome, often results in such changes.

Virus proteases are frequently involved in maturation, although cellular enzymes or a mixture of virus and cellular enzymes are used in some cases. Clearly there is a danger in relying on cellular proteolytic enzymes in that their relative lack of substrate specificity could easily completely degrade the capsid proteins. Virus-encoded proteases on the other hand are usually highly specific for particular amino acid sequences and structures, frequently only cutting one particular peptide bond in a large and complex virus capsid. Moreover, they are often further controlled by being packaged into virus particles during assembly and are only activated when brought into close contact with their target sequence by the conformation of the capsid, by being placed in a local hydrophobic environment, or by changes of pH or metal ion concentration inside the capsid. Retrovirus proteases are good examples of enzymes involved in maturation which are under this tight control. The retrovirus core particle is composed of proteins from the *gag* gene and the protease is packaged into the core prior to its release from the cell on budding. At some stage of the budding process (the exact timing varies for different retroviruses) the protease cleaves the gag protein precursors into the mature products – the capsid, nucleocapsid, and matrix proteins of the mature virus particle (Figure 4.15).

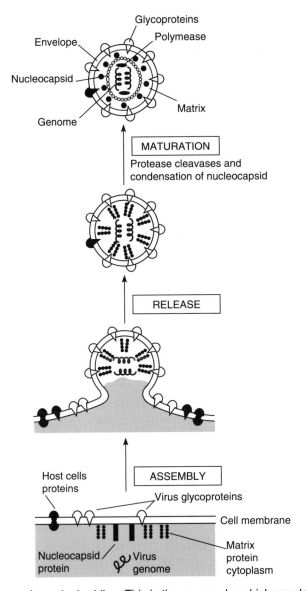

Figure 4.15 Virus release by budding. This is the process by which enveloped virus particles acquire their membranes and associated proteins.

Not all protease cleavage events involved in maturation are this tightly regulated. Native influenza virus hemagglutinin undergoes post-translational modification (glycosylation in the golgi apparatus) and at this stage exhibits receptor-binding activity. However, the protein must be cleaved into two fragments (HA_1 and HA_2) to become able to promote membrane fusion during infection. Cellular trypsin-like enzymes are responsible for this process, which occurs in secretory vesicles as the virus buds into them prior to release at the cell surface. Amantadine and rimantadine are two drugs which are active against influenza A viruses. The action of these closely related agents is complex and incompletely understood, but they are believed to block cellular membrane ion channels. The target for both drugs is the influenza matrix protein (M_2), but resistance to the drug maps to the hemagglutinin gene. The replication of some strains of influenza virus is inhibited at the cell penetration stage and that of others at maturation. The biphasic action of these drugs results from the inability of drug-treated cells to lower the pH of the endosomal compartment (a function normally controlled by the M_2 gene product), and hence to cleave hemagglutinin during maturation. Similarly, retrovirus envelope glycoproteins require cleavage into the surface (SU) and transmembrane (TM) proteins for activity. This process is also carried out by cellular enzymes, but is in general poorly understood.

As already stated, for some viruses, assembly and maturation occur inside the cell and are inseparable, whereas for others, maturation events may occur only after release of the virus particle from the cell. In all cases, the process of maturation prepares the particle for the infection of subsequent cells.

Release

For lytic viruses (such as most non-enveloped viruses), release is a simple process – the infected cell breaks open and releases the virus. Enveloped viruses acquire their lipid membrane as the virus buds out of the cell through the cell membrane or into an intracellular vesicle prior to subsequent release. Virion envelope proteins are picked up during this process as the virus particle is extruded. This process is known as budding. Release of virus particles in this way may be highly damaging to the cell (e.g. rhabdoviruses and herpesviruses) or alternatively may not be (e.g. retroviruses), but in either case, the process is controlled by the virus – the physical interaction of the capsid proteins on the inner surface of the cell membrane forces the particle out through the membrane (Figure 4.15). As mentioned earlier, assembly, maturation, and release are usually simultaneous processes for viruses which are released by budding. The type of membrane from which the virus buds depends on the virus concerned. It can either be the cell surface membrane (retroviruses; togaviruses), the cytoplasmic membrane (orthomyxoviruses; paramyxoviruses; bunyaviruses; coronaviruses; rhabdoviruses; hepadnaviruses), or the nuclear membrane (herpesviruses).

The release of mature virus particles from susceptible host cells by budding presents a problem in that these particles are designed to enter, rather than leave, cells. How do these particles manage to leave the cell surface? In detail, this question remains unanswered, but there are clues as to how the process is achieved. Certain viral envelope proteins are involved in the release phase of replication as well as in the initiating steps. A good example of this is the neuraminidase protein of influenza virus. In addition to being able to reverse the attachment of virus particles to cells via hemagglutinin, neuraminidase is also believed to be important in preventing the aggregation of influenza virus particles and may well have a role in virus release. Apart from specific proteins, viruses which bud have also solved the problem of release by the careful timing of the assembly–maturation–release pathway. Although it may not be possible to separate these stages by means of biochemical analysis, this does not mean that careful spatial separation of these processes has not evolved as a means to solve this problem. Similarly, although we may not understand all the subtleties of the many conformation changes which occur in virus capsids and envelopes during these late stages of replication, virus replication clearly works, in spite of our deficiencies!

Summary

In general terms, virus replication involves three broad stages which are carried out by all types of virus: the initiation of infection, replication and expression of the genome, and finally, the release of mature virions from the infected cell. At a detailed level, there are many differences between the replication processes of different viruses. These are imposed by the biology of prokaryotic or eukaryotic host cells and the nature of the virus genome. Nevertheless, it is possible to derive an overview of virus replication with common stages which, in one form or another, are followed by all viruses.

Further Reading

Baltimore, D. (1971). The expression of animal virus genomes. *Bacteriological Reviews* **35**: 235–41.

Bibliography

Clapham, P.R. *et al.* (1989). Soluble CD4 blocks the infectivity of diverse strains of HIV and SIV for T-cells and monocytes but not for brain and muscle cells. *Nature* **337**: 368–70.

Ellis, E.L. and Delbruck, M. (1934). The growth of bacteriophage. *J. Gen. Physiol.* **22**, 365–84.

Hellen, C.U.T. and Wimmer, E. (1992). The role of proteolytic processing in the morphogenesis of virus particles. *Experientia* **48**: 201–15.

Hershey, A.D. and Chase, M. (1952). Independent functions of viral protein and nucleic acid in growth of bacteriophage. *J. Gen. Physiol.* **26**: 36–56.

Lentz, T.L. (1990). The recognition event between virus and host cell receptor: a target for antiviral agents. *J. Gen. Virol.* **71**: 751–66.

Li, Q.X. *et al.* (1992). Epstein–Barr virus infection and replication in a human epithelial cell line. *Nature* **356**: 347–50.

White, J.M. and Littman, D.R. (1989). Viral receptors of the immunoglobulin superfamily. *Cell* **56**: 725–8.

Chapter 5

Expression

Expression of Genetic Information

The most critical interaction between a virus and its host cell is the requirement by the virus for the cellular apparatus for nucleic acid and protein synthesis. The course of virus replication is determined by tight control of gene expression. There are fundamental differences in the control of these processes in prokaryotic and eukaryotic cells. These differences inevitably affect the viruses that utilize them as hosts. In addition, the relative simplicity and compact size of virus genomes (compared with those of even the simplest cell) imposes further constraints. Cells have evolved varied and complex mechanisms for controlling gene expression by utilizing their extensive genetic capacity. Viruses have had to achieve highly specific quantitative, temporal, and spatial control of expression within much more limited genetic resources. Aspects of this problem have already been discussed in Chapters 3 and 4.

Viruses have counteracted their genetic limitations by the evolution of a range of solutions to these problems. These mechanisms include powerful positive and negative signals which promote or repress gene expression; highly compressed genomes in which overlapping reading frames are commonplace; control signals which are frequently nested within other genes; and several strategies designed to create multiple polypeptides from a single messenger RNA. Gene expression involves regulatory loops mediated by signals which act either in *cis* (affecting the activity of contiguous genetic regions) or in *trans* (giving rise to diffusible products which act on regulatory sites whether or not these are contiguous with the site from which they are produced). For example, transcription promoters are *cis*-acting sequences that are located adjacent to the genes whose transcription they control, while proteins which bind to specific sequences present on any stretch of nucleic acid present in the cell such as 'transcription factors' are examples of *trans*-acting factors. As already mentioned, the (perceived) relative simplicity of virus genomes and the elegance of the control mechanisms which they have evolved feature prominently as models that form the basis of current understanding of genetic regulation.

This chapter assumes familiarity with the mechanisms involved in cellular control of gene expression. However, to illustrate the details of virus gene expression, a very brief summary of some pertinent aspects is given below.

Control of Prokaryote Gene Expression

Bacterial cells are second only to viruses in the specificity and economy of their genetic control mechanisms, and are possibly foremost in terms of the intensity with which these mechanisms have been studied. Genetic control is exercised both at the level of transcription and at subsequent (post-transcriptional) stages of gene expression.

The initiation of transcription is regulated primarily in a negative fashion by the synthesis of *trans*-acting repressor proteins, which bind to operator sequences upstream of protein coding sequences. Collections of metabolically related genes are grouped together and coordinately controlled as 'operons' in this way. Transcription of these operons typically produces a **polycistronic** mRNA which encodes several different proteins. During subsequent stages of expression, transcription is also regulated by a number of mechanisms which act, in Mark Ptashne's famous phrase, as 'genetic switches,' turning on or off the transcription of different genes. Such mechanisms include antitermination, which is controlled by *trans*-acting factors that promote the synthesis of longer transcripts encoding additional genetic information, and by various modifications of RNA polymerase. Bacterial σ (sigma) factors are apoproteins which affect the specificity of the RNA polymerase holoenzyme for different promoters. Several bacteriophages (e.g. phage SP01 of *Bacillus subtilis*) encode proteins which function as alternative σ factors, sequestering RNA polymerase and altering the rate at which phage genes are transcribed. Phage T4 of *E. coli* encodes an enzyme which carries out a covalent modification (ADP-ribosylation) of the host cell RNA polymerase. This is believed to eliminate the requirement of the polymerase holoenzyme for σ factor and to achieve a similar effect to the production of modified σ factors by other bacteriophages.

At a post-transcriptional level, gene expression in bacteria is also regulated by control of translation. The best known viral examples of this phenomenon come from study of bacteriophages of the family *Leviviridae*, such as R17, MS2, and Qβ. In these phages, the secondary structure of the single-stranded RNA phage genome not only regulates the quantities of different phage proteins that are translated, but also operates temporal control of a switch in the ratios between the different proteins produced in infected cells.

Control of Expression in Bacteriophage λ

The genome of phage λ has probably been studied in more detail than any other and illustrates several of the mechanisms described above, including the action of repressor proteins in regulating lysogeny versus lytic replication, and antitermination of transcription by phage-encoded *trans*-acting factors. Such has been the impact of these discoveries that no discussion of the control of virus gene expression could be considered complete without detailed examination of this phage.

Phage λ was discovered at the Pasteur Institute by André Lwoff (who later helped to initiate the idea of operons in bacteria) when he observed that some strains of *E. coli*, when irradiated with ultraviolet light, stopped growing and subsequently lyzed, releasing a crop of bacteriophage particles. Together with

Francois Jaçob and Jaques Monod, he subsequently showed that all the cells of these bacterial strains carried the bacteriophage in a dormant form, known as a prophage, and that the phage could be made to alternate beteween the lysogenic (non-productive) and lytic (productive) growth cycles. After many years of study, our understanding of λ has been refined into a picture which represents one of the best understood and most elegant genetic control systems yet to be investigated.

A simplified genetic map of λ is shown in Figure 5.1. For regulation of the growth cycle of the phage, the structural genes encoding the head and tail components of the virus capsid can be ignored. The pertinent components involved in genetic control are as follows:

- P_L is the promoter responsible for transcription of the left-hand side of the λ genome, including *N* and *cIII*.
- O_L is a short non-coding region of the phage genome (approximately 50 bp) which lies between the *cI* and *N* genes next to P_L.
- P_R is the promoter responsible for transcription of the right-hand side of the λ genome, including *cro*, *cII*, and the genes encoding the structural proteins.
- O_R is a short non-coding region of the phage genome (approximately 50 bp) which lies between the *cI* and *cro* genes next to P_R.
- *cI* is transcribed from its own promoter and encodes a repressor protein of 236 amino acids which binds to O_R, preventing transcription of *cro* but allowing transcription of *cI*, and to O_L, preventing transcription of *N* and the other genes in the left-hand end of the genome.
- *cII* and *cIII* encode activator proteins which bind to the genome, enhancing the transcription of the *cI* gene.
- *cro* encodes a 66 amino acid protein which binds to O_R, blocking binding of the repressor to this site.
- *N* encodes an antiterminator protein which acts as an alternative rho factor for host cell RNA polymerase, modifying its activity and permitting extensive transcription from P_L and P_R.
- *Q* is an antiterminator similar to *N*, but only permits extended transcription from P_R.

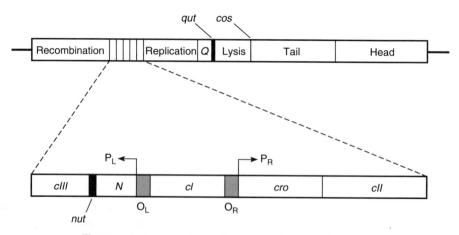

Figure 5.1 Simplified genetic map of bacteriophage λ.

In a newly infected cell, *N* and *cro* are transcribed from P_L and P_R respectively (Figure 5.2). The N protein allows RNA polymerase to transcribe a number of phage genes, including those responsible for DNA recombination and integration of the prophage, as well as *cII* and *cIII*. The N protein acts as a positive transcription

Figure 5.2 Control of expression of the bacteriophage λ genome. A detailed description of the events which occur in a newly infected cell and during lytic infection or lysogeny are given in the text.

regulator. In the absence of the N protein, the RNA polymerase holoenzyme stops at certain sequences located at the end of the N and Q genes, known as the *nut* and *qut* sites, respectively. However, RNA polymerase–N protein complexes are able to overcome this restriction and permit full transcription from P_L and P_R. The RNA polymerase–Q protein complex results in extended transcription from P_R only. As levels of the cII and cIII proteins in the cell build up, transcription of the *cI* repressor gene from its own promoter is turned on.

It is at this point that the critical event occurs which determines the outcome of the infection. The cII protein is constantly degraded by proteases present in the cell. If levels of cII remain below a critical level, transcription from P_R and P_L continues and the phage undergoes a productive replication cycle which culminates in lysis of the cell and the release of phage particles. This is the sequence of events that occurs in the vast majority of cells which become infected. However, in a few rare instances, the concentration of cII protein builds up, transcription of *cI* is enhanced and intracellular levels of the cI repressor protein rise. The repressor binds to O_R and O_L, which prevents transcription of all phage genes (and in particular *cro* – see below) except itself. The level of cI protein is maintained automatically by a negative feedback mechanism, since at high concentrations the repressor also binds to the left-hand end of O_R and prevents transcription of *cI* (Figure 5.3). This autoregulation of cI synthesis keeps the cell in a stable state of lysogeny.

If this is the case, how do such cells ever leave this state and enter a productive, lytic replication cycle? Physiological stress and particularly ultraviolet irradiation of cells results in the induction of a host cell protein, RecA. This protein, whose normal function is to induce the expression of cellular genes which permit the cell to adapt to and survive in altered environmental conditions, cleaves the cI repressor protein. In itself, this would not be sufficient to prevent the cell re-entering the lysogenic state. However, when repressor protein is not bound to O_R, *cro* is transcribed from P_R. This protein also binds to O_R, but unlike cI, which preferentially bind to the right-hand end of O_R, the cro protein binds preferentially to the left-hand end of O_R, preventing the transcription of *cI* and enhancing its own transcription in a positive feedback loop. Thus, the phage is locked into a lytic cycle and cannot return to the lysogenic state.

This description is a highly simplified version of the genetic control of expression in phage λ. A great deal of detail is known about the molecular mechanisms by which the above systems work, but since this could easily fill a whole book on its own, there is insufficient space to recount all of it here. The molecular details of the λ system are not only of isolated interest, but have shaped our understanding of genetic regulation in prokaryotic and also in eukaryotic cells. Determination of the structures of the proteins involved in the above scheme has identified the fundamental principles by which many proteins from unrelated organisms can recognize and bind to specific sequences in DNA molecules. The concepts of proteins with independent DNA-binding and dimerization domains, protein, cooperativity in DNA binding and DNA looping which allows proteins bound at distant sites to interact with one another, have all arisen from the study of λ. It is important to read the references given at the end of this chapter to fully understand the nuances of gene expression in this complex phage, but also to remember the

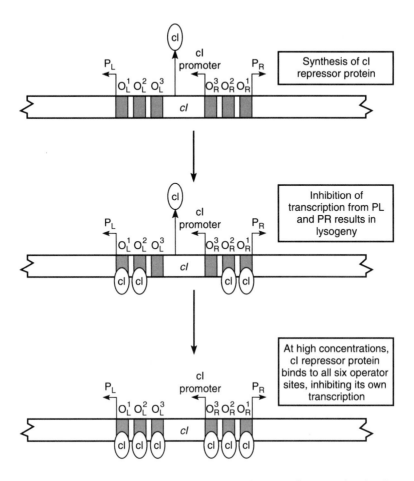

Figure 5.3 Control of lysogeny in bacteriophage λ. See text for details.

design and operation of this system when reading the examples of the regulation of gene expression described in the rest of this chapter.

Control of Eukaryote Gene Expression

Control of gene expression in eukaryotic cells is much more complex than in prokaryotic cells, and involves a 'multilayered' approach in which diverse control mechanisms exert their effects at multiple levels. The first level of control occurs prior to transcription and depends on the local configuration of the DNA. DNA in eukaryotic cells has an elaborate structure, forming complicated and dynamic but far from random complexes with numerous proteins to form **chromatin**. Although the contents of eukaryotic cell nuclei appear amorphous in electron micrographs (at least in interphase), they are actually highly ordered. Chromatin interacts with

Table 5.1 Forms of RNA polymerase in eukaryotic cells

RNA polymerase	Sensitivity to α-amanitin	Cellular genes transcribed	Virus genes transcribed
I	Unaffected	Ribosomal RNAs	—
II	Highly sensitive	Most single-copy genes	Most DNA virus genomes
III	Moderately sensitive	5S rRNA, tRNAs	Adenovirus VA RNAs

the structural backbone of the nucleus, the nuclear matrix, and these interactions are thought to be important in controlling gene expression. Locally, nucleosome configuration and DNA conformation, particularly the formation of left-handed helical 'Z-DNA,' are also important. DNAse I digestion of chromatin does not give an even, uniform digestion pattern, but reveals particular 'DNAse hypersensitive' sites believed to indicate differences in the function of various regions of the chromatin. It is possible, for example, that retroviruses are more likely to integrate into the host cell genome at these sites than elsewhere. Transcriptionally active DNA is also hypomethylated, i.e. there is a relative paucity of nucleotides modified by the covalent attachment of methyl groups in these regions compared with the frequency of methylation in transcriptionally quiescent regions of the genome. The methylation of Moloney murine leukemia virus sequences in pre-implantation mouse embryos has been shown to suppress the transcription of the provirus genome.

The second level of control rests in the process of transcription itself, which again is much more complex than in prokaryotes. There are three forms of RNA polymerase in eukaryotic cells which can be distinguished by their relative sensitivities to the drug α-amanitin and which show specificities for different classes of genes (Table 5.1). The rate at which transcription is initiated is a key control point in eukaryotic gene expression. Initiation is influenced dramatically by sequences upstream of the transcription start site, which function by acting as recognition sites for families of highly specific DNA-binding proteins, known colloquially as 'transcription factors' (Table 5.2). Immediately upstream of the transcription start site is a relatively short region known as the **promoter**. It is at this site that transcription complexes, consisting of RNA polymerase plus accessory proteins, bind to the DNA and transcription begins. However, the sequences further

Table 5.2 Eukaryotic transcription factors

Transcription factor	Molecular weight	Recognition site
SP1	85 000	5'-GGGCGG-3'
NF1 (or 'CTF')	52 000–66 000	5'-GCCAAT-3'
ATF (or 'CREB')	43 000	5'-(T/G)(T/A)CGTCA-3'
AP2	48 000	5'-CCCCAGGG-3'
NFκB	50 000–65 000	5'-GGGACTTTCC-3'

upstream from the promoter influence the efficiency with which transcription complexes form. The rate of initiation depends on the combination of transcription factors bound to these transcription **enhancers**. The properties of these enhancer sequences are remarkable in that they can be inverted and/or moved around relative to the position of the transcription start site without losing their activity and can exert their influence even from a distance of several kilobases away from the start site. This emphasizes the flexibility of DNA, which allows proteins bound at distant sites to interact with one another, as is shown by the protein:protein interactions seen in regulation of phage λ gene expression (above). Transcription of eukaryotic genes results in the production of **monocistronic** mRNAs, each of which is transcribed from its own individual promoter.

At the next stage, gene expression is influenced by the structure of the mRNA produced. The stability of eukaryotic mRNAs varies considerably, some having comparatively long half-lives in the cell (e.g. many hours). The half-lives of others, typically those which encode regulatory proteins, may be very short (e.g. a few minutes). The stability of mRNA depends on the speed with which it is degraded, and this is determined by factors such as its terminal sequences, which consist of a methylated cap structure at the 5′ end and polyadenylic acid at the 3′ end, as well as on the overall secondary structure of the message. However, gene expression is also regulated by differential splicing of heterogeneous (heavy) nuclear RNA (**hnRNA**) precursors in the nucleus, which can alter the genetic 'meaning' of different mRNAs transcribed from the same gene. In eukaryotic cells, control is also exercised during export of RNA from the nucleus to the cytoplasm.

Finally, the process of translation offers further opportunities for control of expression. The efficiency with which different mRNAs are translated varies greatly. These differences result largely from the efficiency with which ribosomes bind to different mRNAs, recognize AUG translation initiation codons in different sequence contexts, and the speed at which different sequences are converted into protein. Certain sequences act as translation enhancers, performing a function analogous to that of transcription enhancers. There are further opportunities to determine the balance between translation of individual open reading frames in individual mRNAs by mechanisms such as ribosomal frameshifting due to the formation of RNA 'pseudoknots,' and termination suppression.

The point of giving this extensive list of eukaryotic gene expression mechanisms is that *all* of them are utilized by viruses to control gene expression. Examples of each type are given in the sections below. If this seems remarkable, it must be remembered that the control of gene expression in eukaryotic cells was unravelled largely by utilizing viruses as model systems. Therefore, finding examples of these mechanisms in viruses is really only a self-fulfilling prophecy.

Genome Coding Strategies

In Chapter 4, genome structure was one element of an arbitrary classification used to divide virus genomes into seven groups (page 103). The other part of such a scheme is the way in which the genetic information of each class of virus genomes is expressed. The replication and expression of virus genomes are inseparably linked, and this is particularly true in the case of RNA viruses. Here, the seven classes of virus genome described in Chapter 4 will be reviewed again, this time examining the way in which the genetic information of each class is expressed.

Class I: Double-stranded DNA (*Adenoviridae; Herpesviridae; Poxviridae; Papovaviridae*)

It was stated in Chapter 4 that this class of virus genomes can be subdivided into two further groups: those in which genome replication is exclusively nuclear (*Adenoviridae, Papovaviridae, Herpesviridae*) and those in which replication occurs in the cytoplasm (*Poxviridae*). In one sense, all of these viruses can be considered as similar, since their genomes all resemble double-stranded cellular DNA, they are essentially transcribed by the same mechanisms as cellular genes. However, there are profound differences between them relating to the degree to which each family is reliant on the host cell machinery.

Papovaviruses

Papovaviruses are heavily dependent on cellular machinery both for replication and gene expression. Polyoma viruses encode *trans*-acting factors (T-antigens) which stimulate transcription (and genome replication). Papillomaviruses in particular are dependent on the cell for replication, which only occurs in terminally differentiated keratinocytes and not in other cell types, although they do encode several *trans*-regulatory proteins.

Adenoviruses

Adenoviruses are also heavily dependent on the cellular apparatus for transcription, but they encode various mechanisms which specifically regulate virus gene expression. These include *trans*-acting transcriptional activators such as the E1A protein and post-transcriptional regulation of expression, which is achieved by alternative splicing of mRNAs and the virus-encoded VA RNAs. Adenovirus infection of cells is divided into two stages, early and late, the latter phase commencing at the time when genome replication occurs. However, in adenoviruses, these phases are less distinct than in herpesviruses (below).

Herpesviruses

These viruses are less reliant on cellular enzymes than the above groups. They encode many enzymes involved in DNA metabolism (e.g. thymidine kinase) and several *trans*-acting factors which regulate the temporal expression of virus genes, controlling the phases of infection. Transcription of the large, complex genome is sequentially regulated in a cascade fashion (Figure 5.4). At least 50 virus-encoded proteins are produced after transcription of the genome by host cell RNA polymerase II. Three distinct classes of mRNAs are made:

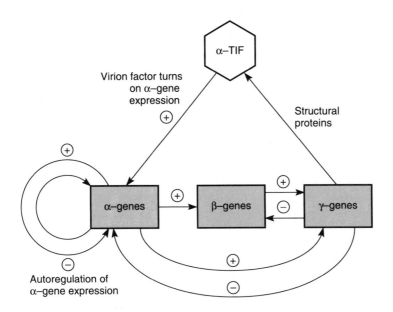

Figure 5.4 Control of expression of the herpesvirus genome. Herpesvirus particles contain a protein called α-TIF (α-gene transcription initiation factor) which turns on α-gene expression in newly infected cells, beginning a cascade of closely regulated events which control the expression of the entire complement of the 70-odd genes in the virus genome.

- α-Immediate-early (IE) mRNAs encode five *trans*-acting regulators of virus transcription;
- β-(Delayed) early mRNAs encode further non-structural regulatory proteins and some minor structural proteins; and
- γ-Late mRNAs encode the major structural proteins.

Gene expression in herpesviruses is tightly and coordinately regulated, as indicated by the following observations (Figure 5.4):

- If translation is blocked shortly after infection (e.g. by treating cells with cycloheximide), early mRNAs immediately accumulate in the nucleus but no further virus mRNAs are transcribed.
- Synthesis of the early gene product turns off the immediate-early products and initiates genome replication.
- Some of the late structural proteins are produced independently of genome replication, whereas others are only produced after replication of the genome from transcription of the newly replicated DNA.

Both the immediate-early and early proteins are required to initiate genome replication. A virus-encoded DNA-dependent DNA polymerase and a DNA-binding protein are involved in genome replication, together with a number of enzymes (e.g. thymidine kinase) which alter cellular biochemistry. The production of all of these proteins is closely controlled.

Poxviruses

Genome replication and gene expression in poxviruses are almost independent of cellular mechanisms (apart from the requirement for host cell ribosomes). Poxvirus genomes encode numerous enzymes involved in DNA metabolism, viral gene transcription, and post-transcriptional modification of mRNAs. Many of these enzymes are packaged within the virus particle (which contains > 100 proteins), enabling transcription and replication of the genome to occur in the cytoplasm (rather than in the nucleus, like all the families described above) almost totally under the control of the virus. Gene expression is carried out by viral enzymes associated with the core of the particle and is divided into two rather indistinct phases:

- *Early genes*: Comprising about 50% of the poxvirus genome, these genes are expressed before genome replication inside a partially uncoated core particle (see Chapter 2), resulting in the production of 5′ capped, 3′ polyadenylated but unspliced mRNAs.
- *Late genes*: These are expressed after genome replication in the cytoplasm, but their expression is also dependent on virus-encoded rather than on cellular transcription proteins (which are located in the nucleus). Like herpesviruses, late gene promoters are dependent on prior DNA replication for activity.

More detailed consideration of some of the mechanisms mentioned above is given later in this chapter (see Transcriptional Control of Expression and Post-transcriptional Control of Expression below).

Class II: Single stranded (+)sense DNA (*Parvoviridae* and *Geminiviruses*)

Both the autonomous and the helper virus-dependent parvoviruses are highly reliant on external assistance for gene expression and genome replication. This is presumably because the very small size of their genomes does not permit them to encode the necessary biochemical apparatus. Thus, they appear to have 'opted' for an extreme form of parasitism, utilizing the normal functions present in the nucleus of their host cells for both expression and replication (Figure 5.5). The geminiviruses of plants also fall into this class of genome structures (see Chapter 3). The expression of their genomes is quite different to that of parvoviruses, but nevertheless is still heavily reliant on host cell functions.

The members of the replication-defective *Dependovirus* genus of the parvovirus family are entirely dependent on adenovirus or herpesvirus superinfection for the provision of further helper functions essential for their replication. The adenovirus genes required as helpers are the early, transcriptional regulatory genes such as E1A, rather than late structural genes, but it has recently been shown that treatment of cells with UV, cycloheximide, or some carcinogens can replace the requirement for helper viruses. Therefore, the help required appears to be for a modification of the cellular environment rather than for a specific virus protein – probably affecting transcription of the defective parvovirus genome.

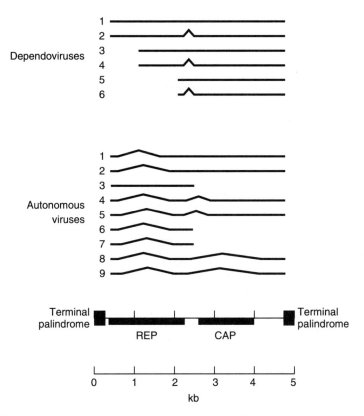

Figure 5.5 Transcription of parvovirus genomes is heavily dependent on host cell factors and results in the synthesis of a series of spliced, subgenomic mRNAs which encode two proteins, rep, which is involved in genome replication and cap, the capsid protein (see text).

Class III: Double-stranded RNA (*Reoviridae*)

All viruses with RNA genomes differ fundamentally from their host cells, which of course possess double-stranded DNA genomes. Therefore, although each virus must be biochemically 'compatible' with its host cell, there are fundamental differences in the mechanisms of virus gene expression from those of the host cell. Reoviruses have multipartite genomes (see Chapter 3) and replicate in the cytoplasm of the host cell. Characteristically for viruses with segmented RNA genomes, on infection, a virus-specific **transcriptase** packaged inside the virus particle produces a separate monocistronic mRNA from each of the genome segments (Figure 5.6). These are translated to produce the non-structural virus proteins required for replication. At later stages of infection, single-stranded (+)sense RNAs are also copied by a virus-encoded **replicase** to produce (−)sense RNA strands, which together form new double-stranded genome segments. After replication of the genome, late mRNAs encoding the structural proteins are transcribed from the newly formed genome segments and are released from a virus core particle to be translated, followed by the formation of mature virus particles.

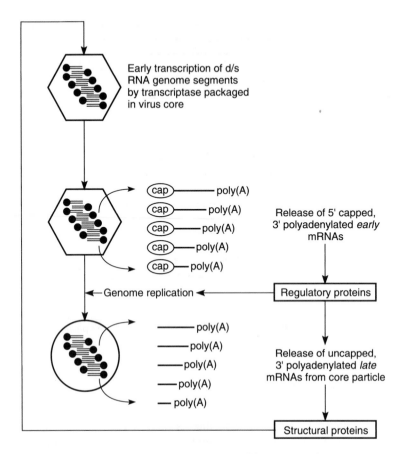

Figure 5.6 Expression of reovirus genomes is initiated by a transcriptase enzyme packaged inside every virus particle. Subsequent events occur in a tightly regulated pattern, the expression of late mRNAs encoding the structural proteins being dependent on prior genome replication.

Class IV: Single-stranded (+)sense RNA

This type of genome occurs in several animal viruses (*Caliciviridae; Coronaviridae; Flaviviridae; Picornaviridae; Togaviridae*) and plant viruses (*Bromoviridae; Carmoviridae; Comoviridae; Cucumoviridae; Dianthoviridae; Furoviridae; Hordeiviridae; Ilarviridae; Luteoviridae; Nepoviridae; Potexviridae; Potyviridae; Sobemoviridae; Tobamoviridae; Tobraviridae; Tombusviridae; Tymoviridae*).

In terms of both the number of different families and the number of individual viruses, this is the largest single class of virus genome. Fundamentally, these virus genomes act as messenger RNAs and are themselves translated immediately after infection of the host cell (see Chapter 3). Not surprisingly with so many representatives, this class of genomes displays a very diverse range of strategies for controlling gene expression and genome replication. However, in very broad terms, the viruses in this class can be subdivided into two groups as follows:

(a) *Production of a **polyprotein** encompassing the whole of the viral genetic information,* which is subsequently cleaved by proteases to produce precursor and mature polypeptides. These cleavages can be a subtle way of regulating the expression of genetic information, alternative cleavages resulting in the production of proteins with quite different properties from a single precursor, e.g. picornaviruses and potyviruses (Figure 5.7). Certain plant viruses with multipartite genomes utilize a very similar strategy for controlling gene expression, although of course, a separate polyprotein is produced from each of the genome segments. The best studied example of this is the comoviruses, whose genome organization is very similar to that of the picornaviruses, and may represent another member of this 'superfamily' (Figure 5.7).

(b) *Production of subgenomic mRNAs,* resulting from two or more rounds of translation of the genome. This strategy is used to achieve temporal separation

Figure 5.7 Positive-sense RNA virus genomes are frequently translated to form a long polyprotein, which is subsequently cleaved by a highly specific virus-encoded protease to form the mature polypeptides.

of what are essentially 'early' and 'late' phases of replication, in which non-structural proteins, including a viral replicase, are produced during the 'early' phase followed by structural proteins in the 'late' phase (Figure 5.8). The proteins produced in each of these phases may result from proteolytic processing of a polyprotein precursor, although this encompasses only part of the virus genome rather than the entire genome, as above. Proteolytic processing offers further opportunities for regulation of the ratio of different polypeptides produced in each phase of replication (e.g. in togaviruses and potexviruses). In addition to proteolysis, some viruses employ another strategy to produce alternate polypeptides from a subgenomic mRNA, either by read-through of a 'leaky' translation stop codon (e.g. tobamoviruses such as TMV), or by deliberate ribosomal frameshifting at a particular site (see Post-transcriptional Control of Expression below).

All of the viruses in this class have evolved mechanisms that allow them to regulate

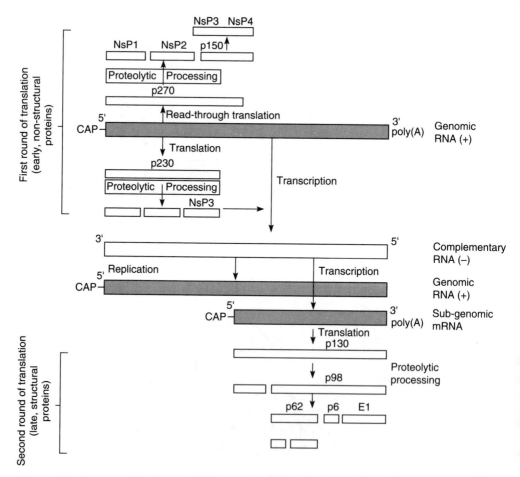

Figure 5.8 Some positive-sense RNA virus genomes (e.g. togaviruses) are expressed by two separate rounds of translation, involving the production of a subgenomic mRNA at a later stage of replication.

their gene expression both in terms of the ratios of different virus-encoded proteins and the stage of the replication cycle when they are produced. Compared with the two classes of DNA virus genomes described above, these mechanisms operate largely independently of those of the host cell. The power and flexibility of these strategies is reflected very clearly in the overall success of the viruses in this class, as determined by the number of different representatives known and the number of different hosts which they infect.

Class V: Single-stranded (−)sense RNA (*Orthomyxoviridae*; *Paramyxoviridae*; *Rhabdoviridae*; *Filoviridae*; *Bunyaviridae*)

As discussed in Chapter 3, the genomes of these viruses may be either segmented or non-segmented.

The first step in the replication of segmented orthomyxovirus genomes is transcription of the (−)sense vRNA by the virion-associated RNA-dependent RNA polymerase to produce (predominantly) monocistronic mRNAs, which also serve as the template for subsequent genome replication (Figure 5.9). As with all (−)sense RNA viruses, packaging of this virus-specific transcriptase/replicase within the virus nucleocapsid is essential, since no host cell contains any enzyme capable of decoding and copying the RNA genome.

In the other families, which have non-segmented genomes, monocistronic mRNAs are also produced. Here, however, these messages must be produced from a single, long (−)sense RNA molecule. Exactly how this is achieved is not clear. It is possible that a single, genome-length transcript is cleaved after transcription to form the separate mRNAs, but it is more likely that these are produced individually by a stop and start mechanism of transcription regulated by the conserved intergenic sequences present between each of the virus genes (see Chapter 3). On the surface, such a scheme of gene expression might appear to offer few opportunities for regulation of the relative amounts of different virus proteins. If this were true, this would be a major disadvantage, since all viruses require far more copies of the

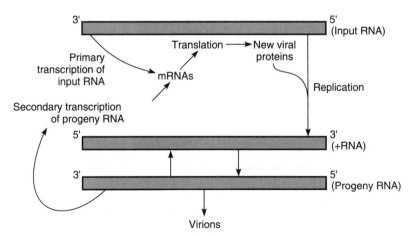

Figure 5.9 Generalized scheme for the expression of negative-sense RNA virus genomes.

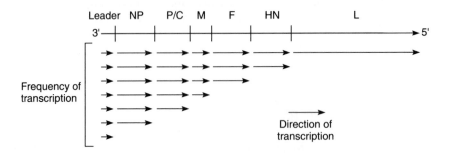

Figure 5.10 Paramyxovirus genomes exhibit 'transcriptional polarity.' Transcripts of genes at the 3′ end of the virus genome are more abundant than those of genes at the 5′ end of the genome, permitting regulation of the relative amounts of structural (3′ genes) and non-structural (5′ genes) proteins produced.

structural proteins (e.g. nucleocapsid protein) than of the non-structural proteins (e.g. polymerase) for each virion produced. Clearly, in practice the ratio of different proteins is regulated both during transcription and afterwards. In paramyxoviruses for example, there is a clear polarity of transcription from the 3′ end of the virus genome to the 5′ end, which results in the synthesis of far more mRNAs for the structural proteins encoded in the 3′ end of the genome than for the non-structural proteins located at the 5′ end (Figure 5.10). Similarly, the advantage of producing monocistronic mRNAs is that the translational efficiency of each message can be varied with respect to the others (see Post-transcriptional Control of Expression below).

Class VI: Single-stranded (+)sense RNA with DNA intermediate (*Retroviridae*)

The retroviruses are perhaps the ultimate case of reliance on host cell transcription machinery. The RNA genome does not serve as mRNA, but as a template for reverse transcription into DNA (see Chapter 3). Once integrated into the host cell genome, the DNA provirus is, to all intents and purposes, under the control of the host cell, and is transcribed exactly as are other cellular genes. Some retroviruses, however, have evolved a number of transcriptional and post-transcriptional mechanisms which allow them to control the expression of their genetic information and these are discussed in detail in the relevant sections later in this chapter.

Class VII: Double-stranded DNA with RNA intermediate (*Hepadnaviridae*; *Caulimovirus*)

Expression of the genomes of these viruses is complex and relatively poorly understood. The hepadnaviruses contain a number of overlapping reading frames clearly designed to squeeze as much coding information as possible into a compact genome. The X gene encodes a transcriptional *trans*-activator, but its role in infection is unknown. At least two mRNAs are produced from independent promoters, each of which encodes several proteins and the larger of which is also the template for reverse transcription during the formation of the virus particle (see Chapter 3).

Expression of caulimovirus genomes is similarly complex, although there appear to be similarities with hepadnaviruses in that two major transcripts are produced, each of which encodes several polypeptides and the larger of which is the template for reverse transcription during the formation of the virus genome.

Transcriptional Control of Expression

Having reviewed the general strategies used by different groups of viruses to regulate gene expression, the rest of this chapter will concentrate on more detailed explanations of specific examples from some of the viruses mentioned above, beginning with control of transcription in SV40, a member of the *Papovaviridae* family.

Few other genomes, viral or cellular, have been studied in such detail as that of SV40, which has been a paradigm for the study of eukaryotic transcription mechanisms (and particularly, DNA replication – see Chapter 6) for many years. In this sense, SV40 provides a eukaryotic parallel with the importance of the phage λ genome in understanding prokaryotic gene expression. *In vitro* systems exist for both transcription and replication of the SV40 genome and it is believed that all of the viral and cellular DNA-binding proteins involved in both of these processes are known. The SV40 genome encodes two T-antigens ('tumour antigens') known as large T-antigen and small T-antigen after the sizes of the proteins (Figure 5.11). Replication of the double-stranded DNA genome of SV40 occurs in the nucleus of the host cell. Transcription of the genome is carried out by host cell RNA polymerase II and large T-antigen plays a vital role in regulating transcription of the virus genome. Small T-antigen is not essential for virus replication, but allows virus DNA to accumulate in the nucleus. Both proteins contain 'nuclear localization signals' which results in their accumulation in the nucleus, where they migrate after being synthesized in the cytoplasm.

Soon after infection of permissive cells, early mRNAs are expressed from the early promoter, which contains a strong transcription enhancer element (the 72 bp sequence repeats), allowing it to be active in newly infected cells (Figure 5.12). The early proteins synthesized are the two T-antigens. As the concentration of large T-antigen builds up in the nucleus, transcription of the early genes is repressed by direct binding of the protein to the origin region of the virus genome, preventing transcription from the early promoter and causing the switch to the late phase of infection. As already mentioned, large T-antigen is also required for replication of the genome, which is considered further in Chapter 7. After DNA replication has occurred, transcription of the late genes occurs from the late promoter, and results in the synthesis of the structural proteins, VP1, VP2, and VP3. Therefore, the role of the SV40 T-antigen in controlling the transcription of the genome is akin to that

Figure 5.11 Organization and protein-coding potential of the SV40 genome.

Figure 5.12 Control of transcription of the SV40 genome.

of a 'switch' and readers should compare the functioning of this system with the description of bacteriophage λ gene expression control given earlier (page 111).

Another area where control of viral transcription has received much attention over the last few years are the human retroviruses, human T-cell leukemia virus (HTLV) and human immunodeficiency virus (HIV). Integrated DNA proviruses are formed by reverse transcription of the RNA retrovirus genome, as described in Chapter 3. The presence of numerous binding sites for cellular transcription factors in the long terminal repeats (LTRs) of these viruses have been analyzed by 'DNAse

I footprinting' and 'gel-shift' assays (Figure 5.13). Together, the 'distal' elements (such as NKκB and SP1 binding-sites) and 'proximal' elements (such as the TATA box) make up a functional transcription promoter in the U3 region of the LTR (see Chapter 3). However, the basal activity of these promoters on their own is relatively weak, and only results in limited transcription of the provirus genome by RNA polymerase II.

Both HTLV and HIV encode proteins which are *trans*-acting positive regulators of transcription, the tax protein of HTLV and the HIV tat protein (Figure 5.14). These proteins act to increase transcription from the viral LTR by a factor of at least 50–100 times that of the basal rate from the 'unaided' promoter. The mechanism by which these proteins function is not clear at the present time. Unlike T-antigen and the early promoter of SV40, neither the tax nor the tat protein (which have no structural similarity to one another) bind directly to the LTR. The HIV tat protein binds to a stem-loop structure at the 5′ end of mRNAs transcribed from the LTR, known as the 'TAR' (Trans-Activation Response) element. It has been suggested that tat and TAR might function as antiterminators of transcription in an analogous way to the N protein/*nut* site in phage λ (see page 112). However, this is not certain, and it is possible that tat (and tax) might act, either directly or indirectly, by somehow modifying the transcription initiation complex which is assembled at the promoter site in the viral LTR. More work is necessary to decide which, if any, of these potential explanations is closest to the truth.

The HTLV tax and HIV tat proteins are positive regulators of the basal promoter in the provirus LTR, and are under the control of the virus, since synthesis of these proteins is dependent on the promoters which they themselves activate (Figure 5.15).

Figure 5.13 Cellular transcription factors which interact with retrovirus LTRs. These cellular DNA-binding proteins are involved in regulating both the basal and *trans*-activated levels of transcription from the promoter in the U3 region of the LTR.

Figure 5.14 Expression of the HTLV and HIV genomes.

On its own, this would be an unsustainable system, since it would result in unregulated positive feedback, which might be acceptable in a lytic replication cycle, but would not be appropriate for a retrovirus integrated into the genome of the host cell. Therefore, these viruses each encode an additional protein, the rex and rev proteins from HTLV and HIV respectively, which further regulate gene expression at a post-transcriptional level (see Post-transcriptional Control of Expression).

In general terms, it can be stated that control of transcription is a critical step in viral replication and in all cases, is closely regulated. Even some of the simplest virus genomes, such as SV40, encode proteins which regulate their transcription. Many virus genomes encode *trans*-acting factors which modify and/or direct the cellular transcription apparatus. Examples of this include HTLV and HIV as described above, but also the X protein of hepadnaviruses, rep protein of parvoviruses, E1A protein of adenoviruses (see below) and the immediate early proteins of herpesviruses. The expression of RNA virus genomes is similarly tightly controlled, but this process is carried out by virus-encoded transcriptases and has been less intensively studied and is generally much less well understood than transcription of DNA genomes.

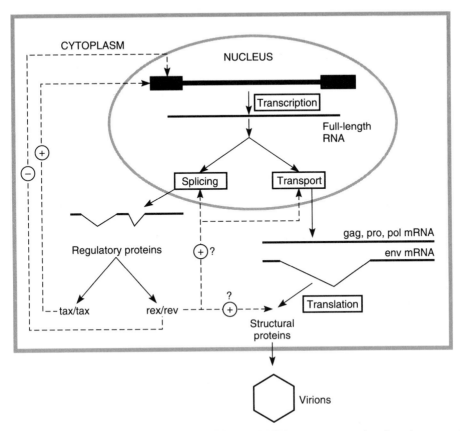

Figure 5.15 *Trans*-acting regulation of HTLV and HIV gene expression by virus-encoded proteins. The tax and tat proteins (from HTLV and HIV respectively) act at a transcriptional level and stimulate the expression of all virus genes. The rex (HTLV) and rev (HIV) proteins act post-transcriptionally and regulate the balance of expression between virion proteins and regulatory proteins.

Post-transcriptional Control of Expression

In addition to control of the process of transcription itself, the expression of viral genetic information is also governed at a number of additional stages between the formation of the primary RNA transcript and the completion of the finished polypeptide. There are many generalized, subtle controls, such as the differential stability of various mRNAs, which are undoubtedly employed by viruses to regulate the flow of genetic information from their genomes into proteins. This section, however, will only describe a few well-researched, specific examples of post-transcriptional regulation.

Many DNA viruses which replicate in the nucleus encode mRNAs which must be spliced by cellular mechanisms to remove intervening sequences (introns) before being translated. This type of modification only applies to viruses which replicate in the nucleus (and not, for example, to poxviruses) since it requires the processing of mRNAs by nuclear apparatus before they are transported into the cytoplasm for translation. However, several virus families have taken advantage of this capacity of their host cells in order to compress more genetic information into their genomes. Some of the best examples of such a reliance on **splicing** are the parvoviruses, transcription of which results in multiple spliced, polyadenylated transcripts in the cytoplasm of infected cells, enabling them to produce multiple proteins from their 5 kb genomes (Figure 5.5), and similarly, papovaviruses such as SV40 (Figure 5.11). In contrast, the large genetic capacity of herpesviruses makes it possible for these viruses to produce mostly unspliced monocistronic mRNAs, each of which is expressed from its own promoter, and hence renders extensive splicing to produce the required repertoire of proteins unnecessary.

One of the best studied examples of the splicing of viral mRNAs is the expression of the adenovirus genome (Figure 5.16). Several 'families' of adenovirus genes are expressed via differential splicing of precursor hnRNA transcripts. This is particularly true for the early genes which encode *trans*-acting regulatory proteins expressed immediately after infection. The first proteins to be expressed, E1A and E1B, are encoded by a transcriptional unit on the *r*-strand at the extreme left-hand end of the adenovirus genome (Figure 5.16). These proteins are primarily transcriptional *trans*-regulatory proteins akin to the tax and tat proteins described above, but are also involved in transformation of adenovirus-infected cells (see Chapter 6). Five polyadenylated, spliced mRNAs are produced (13S, 12S, 11S, 10S, 9S) which encode five related E1A polypeptides (containing 289, 243, 217, 171, and 55 amino acids respectively) (Figure 5.17). All of these proteins are translated from the same reading frame and have the same amino and carboxyl termini. The differences between them are a consequence of differential splicing of the E1A transcriptional

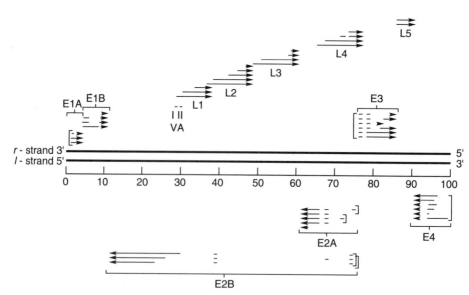

Figure 5.16 Transcription of the adenovirus genome. The arrows show the position of exons in the virus genome which are joined by splicing to produce families of virus proteins.

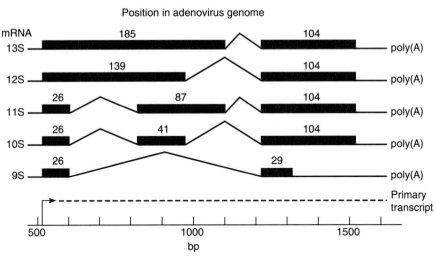

Position in adenovirus genome

Numbers shown above each box are the number of amino acids encoded by each exon

Figure 5.17 Expression of the adenovirus E1A proteins.

unit and result in major differences in their functions. The 289 and 243 amino acid peptides are transcriptional activators. Although these proteins activate transcription from all of the early adenovirus promoters, it has been discovered that they also seem to be 'promiscuous,' activating many RNA polymerase II-responsive promoters which have so far been examined. There are no obvious common sequences present in all of these promoters and there is no evidence that the E1A proteins bind directly to DNA. It is believed that E1A functions indirectly, either by chemically modifying a cellular transcription factor or by binding to the promoter via protein:protein interactions with an intermediate DNA-binding protein. The 217, 171, and 55 amino acid proteins are not active as transcriptional *trans*-activators, but are involved in cell transformation and possibly in regulation of expression.

Synthesis of E1A starts a cascade of transcriptional activation by turning on transcription of the other adenovirus early genes, E1B, E2, E3, and E4 (Figure 5.16). After the virus genome has been replicated, this cascade eventually results in transcription of the late genes encoding the structural proteins. The transcription of the E1A itself is a balanced, self-regulating system. The immediate early genes of DNA viruses typically have strong enhancer elements upstream of their promoters. This is because in a newly infected cell, there are no virus proteins present and the enhancer is required to 'kick-start' expression of the virus genome. The immediate early proteins synthesized are transcriptional activators which turn on expression of other virus genes, and E1A functions in exactly this way. However, although E1A *trans*-activates its own promoter, the protein represses the function of the upstream enhancer element and therefore at high concentrations, also downregulates its own expression (Figure 5.18).

Early gene expression:
Positive regulation

Late gene expression:
Negative regulation

⊠ Enhancer element ◆ Promoters → RNA transcripts

Figure 5.18 Regulation of adenovirus gene expression.

The next stage at which expression can be regulated is during export of mRNA from the nucleus to the cytoplasm. Once again, the best studied example of this phenomenon comes from the adenovirus family. The VA genes encode two small RNAs transcribed from the *r*-strand of the genome by RNA polymerase III (whose normal function is to transcribe similar small RNAs such as 5S ribosomal RNA and tRNAs) during the late phase of virus replication (Figure 5.16). Both VA RNA I and VA RNA II have a high degree of secondary structure and neither molecule encodes any polypeptide – in these two respects they are similar to tRNAs. The way in which these two RNAs act is not completely understood, but their net effect is to boost the synthesis of adenovirus late proteins. The VA RNAs are found to be associated with the small nuclear ribonucleoprotein complexes (snRNPs) which are part of the nuclear splicing apparatus, but there is also evidence that VA RNA I selectively promotes the translation of adenovirus mRNAs in the following way. Virus infection of cells stimulates the production of interferons (see Chapter 6). One of the actions of interferons is to activate a cellular protein kinase, known as DAI or p68, which inhibits the initiation of translation. VA RNA I binds to this kinase, preventing its activity and therefore relieves the inhibition on translation. As discussed in Chapter 6, the effects of interferons on the cell are generalized and result in inhibition of the translation of both cellular and viral mRNAs. VA RNA I may be able to promote selectively the translation of adenovirus mRNAs at the expense of cellular mRNAs whose translation remains inhibited.

The HTLV rex and HIV rev proteins mentioned earlier also act to promote the selective translation of specific virus mRNAs. These proteins regulate the differential expression of the virus genome but do not, so far as is known, substantially alter the expression of cellular mRNAs. Both of these proteins appear to function in a similar way, and although not related to one another in terms of their amino acid sequences, the HTLV rex protein has been shown to be able to substitute functionally for the HIV rev protein. Certain negative-regulatory sequences in the HIV and HTLV genomes appear to cause the retention of virus mRNAs in the nucleus of the infected cell. These sequences are located in the intron regions which are removed from spliced mRNAs encoding the tax/tat and rex/rev proteins (Figure 5.14). Therefore, immediately after infection, these proteins are expressed. Tax and tat stimulate enhanced transcription from the virus LTR (Figure 5.15). However, unspliced or singly spliced mRNAs encoding the *gag*, *pol*, and *env* gene products are only expressed when sufficient rex/rev protein is present in the cell. Both proteins bind to a region of secondary structure formed by a particular sequence in the mRNA. Whether, after binding, they act in the nucleus by controlling the transport of mRNA from the nucleus to the cytoplasm, or alternatively, in the cytoplasm, promoting translation of mRNA by enhancing the formation or activity of polyribosomes, is not yet clear.

The efficiency with which different mRNAs are translated varies considerably. This is determined by a number of factors, including the stability and secondary structure of the RNA, but the main one appears to be the particular nucleotide sequence surrounding the AUG translation initiation codon which is recognized by ribosomes. The most favorable sequence for initiation is GCC(A/G)CC*AUG*GG, although there can be considerable variation within this sequence. A number of viruses use variations of this sequence to regulate the amounts of protein synthesized from a single mRNA. An example of this are the tax and rex proteins of HTLV, which are encoded by overlapping reading frames in the same doubly spliced 2.1 kb mRNA (Figure 5.14). The AUG initiation codon for the rex protein is upstream of that for tax, but provides a less favorable context for initiation of translation than the sequence surrounding the tax AUG codon. This is known as the 'leaky scanning' mechanism, since it is believed that the ribosomes scan along the mRNA before initiating translation. Therefore, the relative abundance of rex protein in HTLV-infected cells is considerably less than that of the tax protein, even though both are encoded by the same mRNA.

Picornavirus genomes illustrate an alternative mechanism for controlling the initiation of translation. Although these genomes are genetically economical (i.e. have discarded most *cis*-acting control elements and express their entire coding capacity as a single polyprotein), they have retained very long non-coding regions at their 5′ ends (5′ NCR), comprising approximately 10% of the entire genome. These sequences are involved in the replication and possibly the packaging of the virus genome. Translation of most cellular mRNAs is believed to be initiated as described above, i.e. ribosomes recognize the 5′ end of the mRNA and scan along the nucleotide sequence until they reach an AUG initiation codon, when translation begins. Picornavirus genomes are not translated in this way. The 5′ end of the RNA is not capped and thus is not recognized by ribosomes in the same way as other mRNAs, but is modified by the addition of the VPg protein (see Chapters 3 and 6). Additionally, there are multiple AUG codons in the 5′ NCR upstream of the start of the polyprotein coding sequences which are not recognized by ribosomes. Several experiments have shown that a different mechanism for controlling the initiation of translation is used by these viruses. Translation of artificially mutated picornavirus mRNAs *in vitro* and the construction of bicistronic picornavirus genomes bearing additional 5' NCR signals in the middle of the polyprotein have resulted in the concept of the ribosome 'landing pad,' or internal ribosomal entry site (IRES). Rather than scanning along the RNA from the 5' end, it appears that ribosomes recognize specific internal sequences in the picornavirus 5' NCR which signal them to bind to the RNA and begin translation. This is a precise method for controlling the translation of virus proteins. Very few cellular mRNAs utilize this mechanism and it is not yet known whether it is employed by other viruses.

Many viruses belonging to different families compress their genetic information by encoding different polypeptides in overlapping reading frames. The problem with this strategy lies in decoding the information. If each polypeptide is expressed from a monocistronic mRNA transcribed from its own promoter, the additional *cis*-acting sequences required to control and coordinate expression might cancel out any genetic advantage gained. More importantly, there is the problem of coordinately regulating the transcription and translation of multiple different messages. Therefore, it is highly desirable to express several polypeptides from a single RNA transcript, and the examples described above illustrate several mechanisms by which this can be achieved, namely, differential splicing and control of RNA export from the nucleus or initiation of translation.

There is, however, an additional mechanism known as ribosomal frameshifting which is used by several groups of viruses to achieve the same end. The best studied examples of this phenomenon come from retrovirus genomes, but coronaviruses and some plant viruses from the tymovirus group also use a similar mechanism. (Such frameshifting is also known to occur during the expression of certain bacterial genes but to date, has not been observed in eukaryotic cells other than during virus infection.) Retrovirus genomes are transcribed to produce at least two 5' capped, 3' polyadenylated mRNAs. Spliced mRNAs encode the envelope proteins and, in more complex retroviruses such as HTLV and HIV, additional proteins such as tax/tat and rex/rev proteins (Figure 5.14). A long, unspliced transcript encodes the *gag*, *pro*, and *pol* genes and also forms the genomic RNA which is packaged into virions. The problem faced by retroviruses is how to express three different proteins from one long transcript. The arrangement of the three genes varies in different viruses. In some cases (e.g. HTLV), they occupy three different reading frames, while in others (e.g. HIV), the protease (*pro*) gene forms an extension at the 5' end of the *pol* gene (Figure 5.19). In the latter case, the protease and polymerase (i.e. reverse transcriptase) are expressed as a polyprotein which is autocatalytically cleaved into the mature proteins in a process which is similar to the cleavage of picornavirus polyproteins.

At the boundary between each of the three genes there is a particular sequence which usually consists of a tract of reiterated nucleotides, such as UUUAAAC (Figure 5.20). Remarkably, this sequence is rarely encountered in protein-coding sequences and therefore appears to be specifically used for this type of regulation. Most ribosomes encountering this sequence will translate it without difficulty and continue on along the transcript until a translation stop codon is reached. However, a proportion of the ribosomes which attempt to translate this sequence slip back by one nucleotide before continuing to translate the message, but now in a different

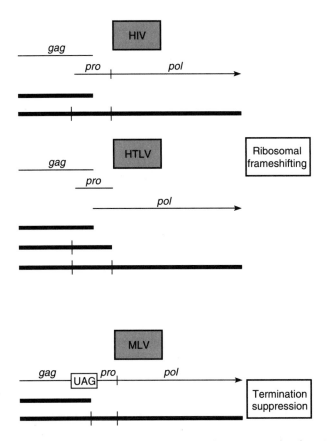

Figure 5.19 Ribosomal frameshifting and termination suppression in retroviruses.

(i.e. −1) reading frame. Because of this, the UUUAAAC sequence has been termed the 'slippery sequence,' and the result of this −1 frameshifting is the translation of a polyprotein containing alternative information from a different reading frame. This mechanism also allows the virus to control the ratios of the proteins produced. Since only a proportion of ribosomes undergo frameshifting at each slippery sequence, there is a gradient of translation from the reading frames at the 5′ end of the mRNA to those at the 3′ end.

However, the slippery sequence alone only results in a low frequency of frameshifting, which appears to be inadequate to produce the amount of protease and reverse transcriptase protein required by the virus. Therefore, there are additional sequences which further regulate this system and increase the frequency of frameshift events. A short distance downstream of the slippery sequence there is an inverted repeat which allows the formation of a stem-loop structure in the mRNA (Figure 5.20). A little further on, there is an additional sequence which is complementary to the nucleotides in the loop and which allows base-pairing between these two regions of the RNA. The net result of this combination of sequences is the formation of what is known as an RNA 'pseudoknot.' This secondary structure in the mRNA causes ribosomes translating the message to

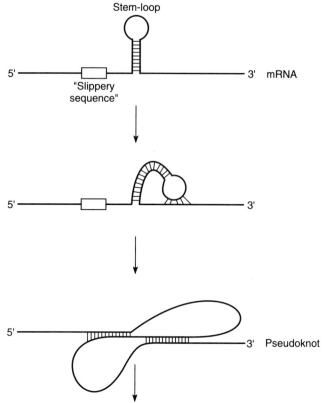

Figure 5.20 RNA pseudoknot formation – the mechanism by which ribosomal frameshifting occurs.

pause at the position of the slippery sequence upstream, and this slowing or pausing of the ribosome during translation increases the frequency at which frameshifting occurs, thus boosting the relative amounts of the proteins encoded by the downstream reading frames. It is easy to imagine how this system can be fine-tuned by subtle mutations which alter the stability of the pseudoknot structure and thus the relative expression of the different genes.

The final method of translational control to be considered is termination suppression. This is a mechanism similar in many respects to frameshifting, and permits multiple polypeptides to be expressed from individual reading frames in a single mRNA. In some retroviruses, such as murine leukemia virus (MLV), the *pro* gene is separated from the *gag* gene by a UAG termination codon rather than a 'slippery sequence' and pseudoknot (Figure 5.19). In the majority of cases, translation of MLV mRNA terminates at this sequence, giving rise to the gag proteins. However, in a few instances, the UAG stop codon is suppressed and translation continues, producing a gag–pro–pol polyprotein, which subsequently cleaves itself to produce the mature proteins. The overall effect of this system is much the same as ribosomal frameshifting, the relative ratios of gag and pro/pol proteins being controlled by the frequency with which ribosomes traverse or terminate at the UAG stop codon.

Summary

Control of gene expression is a vital element of virus replication. Coordinate expression of groups of virus genes results in successive phases of gene expression, typically, immediate early genes which encode 'activator' proteins, early genes which encode further regulatory proteins and late genes which encode virus structural proteins. Viruses make use of the biochemical apparatus of their host cells to express their genetic information as proteins and, consequently, utilize the appropriate biochemical language recognized by the cell; for example, viruses of prokaryotes produce polycistronic mRNAs, while viruses with eukaryotic hosts produce mostly monocistronic mRNAs. Some viruses of eukaryotes do produce polycistronic mRNA in order to assist with the coordinate regulation of multiple genes. In addition, viruses rely on specific *cis-* and *trans-*acting mechanisms to manipulate the biology of their host cells and to enhance and coordinate the expression of their own genetic information.

Further Reading

Adams, R.L.P. (1986). *The Biochemistry of the Nucleic Acids* (10th edition). Chapman and Hall, New York.

Alberts, B., Bray, D., Lewis, J., Raff, M., Roberts, K. and Watson, J.D. (Eds) (1989). Control of gene expression. In *Molecular Biology of the Cell* (2nd edition) Garland, New York, chap. 10.

Ptashne, M. (1986). *A Genetic Switch.* Cell Press/Blackwell Scientific, Oxford.

Bibliography

Boulanger, P.A. and Blair, G.E. (1991). Expression and interactions of human adenovirus oncoproteins. *Biochem. J.* **275**: 281–99.

Gaynor, R. (1992). Cellular transcription factors involved in the regulation of HIV-1 gene expression. *AIDS* **6**: 347–63.

Kozak, M. (1991). Structural features in eukaryotic messenger RNAs that modulate the initiation of translation. *J. Biol. Chem.* **266**: 19867–70.

Ptashne, M. (1986). Gene regulation by proteins acting nearby and at a distance. *Nature* **322**: 697–701.

Chapter 6

Infection

Virus infection of higher organisms is the cumulative result of all the processes of replication and gene expression described in earlier chapters. When blended together, these determine the overall course of each infection. Virus infections range in complexity and duration from a very brief, superficial interaction between the virus and its host, to infections which may encompass the entire life of the host organism, from before birth to its eventual death, and in which many different tissues and organs are infected. One of the most common misconceptions is that virus infection inevitably results in disease. In reality, the converse is true – only a small minority of virus infections give rise to any disease symptoms.

This chapter provides an overview of the numerous patterns of virus infection and forms an introduction to the consideration of viral pathogenesis in Chapter 7. Most of this chapter is concerned with the infection of vertebrate animals by viruses (following a short section on plants). The infection of bacteria by bacteriophages will not be specifically considered here, although there are parallels between these infections and some virus infections of vertebrates. Unlike previous and subsequent chapters, this chapter deals primarily with the interaction of viruses with intact organisms, rather than with the molecular biologist's usual concern about the interaction between a virus and the cell.

Virus Infections of Plants

The overall course of virus replication is determined by a dynamic interaction between the virus and its host organism. Clearly, therefore, there are major differences between virus infections of plants and those of vertebrates. The study of plant viruses is a highly specialist area, and readers who are particularly interested in these viruses would be well advised to consult more specialist publications dealing specifically with this topic. However, because of its general interest, because of the comparisons that can be made with virus infections of vertebrates, and not least because of its tremendous economic importance to agriculture, this chapter begins with a brief consideration of virus infections of plants.

The mechanism by which plant viruses are transmitted between hosts is of considerable importance. In economic terms, viruses are only of importance if it is likely that they will spread to crops during their commercial lifetime, which of course varies greatly between very short extremes in horticultural production and very long extremes in forestry. There are a number of routes by which plant viruses may be transmitted:

- *Seeds*: These may transmit virus infection either due to external contamination of the seed with virus particles, or due to infection of the living tissues of the embryo. Transmission by this route leads to early outbreaks of disease in new crops which are usually initially focal in distribution, but may subsequently be transmitted to the remainder of the crop by other mechanisms (below).
- *Vegetative propagation/grafting*: These techniques are cheap and easy methods of plant propagation, but provide the ideal opportunity for viruses to spread to new plants.
- *Vectors*: Many different groups of living organisms can act as vectors and spread viruses from one plant to another:
  - Bacteria (e.g. *Agrobacterium tumefaciens* – the Ti plasmid of this organism has been used experimentally to transmit virus genomes between plants)
  - Fungi
  - Nematodes
  - Arthropods: Insects – aphids, leafhoppers, planthoppers, beetles, etc.
    Arachnids – mites
- *Mechanical*: Mechanical transmission of viruses is the most widely used method for experimental infection of plants and is usually achieved by rubbing virus-containing preparations into the leaves, which in most plant species are particularly susceptible to infection. However, this is also an important natural method of transmission. Virus particles may contaminate soil for long periods and may be transmitted to the leaves of new host plants as wind-blown dust or as rain-splashed mud.

Transmission of plant viruses by insects is of particular agricultural importance. Extensive areas of monoculture and the inappropriate use of pesticides which kill natural predators can result in massive population booms of insects such as aphids. Transfer by insect vectors is a particularly efficient means of virus transmission. In some instances, viruses are transmitted mechanically from one plant to the next by the vector and the insect is merely a means of distribution, flying or being carried on the wind for long distances (sometimes hundreds of miles). Insects which bite or suck plant tissues are, of course, the ideal means of transmitting viruses to new hosts. This is known as non-propagative transmission. However, in other cases (e.g. many plant rhabdoviruses) the virus may also infect and multiply in the tissues of the insect (**propagative transmission**) as well as those of host plants. In these cases, the vector serves as a means not only of distributing the virus, but also of amplifying the infection.

Initially, most plant viruses multiply at the site of infection, giving rise to localized symptoms such as necrotic spots on the leaves. Subsequently, the virus may be distributed to all parts of the plant either by direct cell-to-cell spread or by the vascular system, giving rise to a **systemic infection** involving the whole plant. Typically, this might result in effects such as growth retardation, mosaic patterning on the leaves, yellowing, wilting, etc. These macroscopic symptoms result from:

necrosis of cells, caused by direct damage due to virus replication; **hypoplasia**, i.e. localized retarded growth frequently leading to mosaicism (the appearance of thinner, yellow areas on the leaves); or **hyperplasia**, which is excessive cell division or the growth of abnormally large cells, resulting in the production of swollen or distorted areas of the plant.

Plants might be seen as 'sitting targets' for virus infection – unlike animals, they cannot run away. However, plants exhibit a range of responses to virus infections designed to minimize their effects. Initially, infection results in a 'hypersensitive response,' which is manifested as the synthesis of a range of new proteins, the 'PR proteins.' Although this system is poorly understood, at least some of these proteins have been characterized and have been shown to be proteases, which presumably destroy virus proteins, limiting the spread of the infection. There is some similarity here between this response and the production of interferons by animals (see below). In addition, systemic resistance to virus infection is a naturally occurring phenomenon in some strains of plant. This is clearly a highly desirable characteristic and is highly prized by plant breeders, who try to spread this attribute to economically valuable crop strains. There are probably many different mechanisms involved in systemic resistance, but in general terms, there is a tendency towards increased local necrosis as substances such as proteases and peroxidases are produced by the plant to destroy the virus and to prevent its spread and subsequent systemic disease. Recently, virus-resistant plants have been created by the production of transgenic plants expressing certain virus proteins. These proteins interfere with virus replication without producing the pathogenic consequences of infection. One such mechanism is the synthesis of virus coat proteins, which swamp the receptors on the host cells and do not permit infection by the virus. This is a very promising technology which offers the possibility of substantial increases in agricultural production without the use of expensive, toxic, and ecologically damaging chemicals (fertilizers, herbicides, or pesticides), but is, at present, still in its infancy.

Immune Responses to Virus Infections in Animals

The most significant response to virus infection in vertebrates is the activation of both the cellular and humoral arms of the immune system. Clearly, a thorough description of all the events involved in the immune response to the presence of foreign antigens in the body is beyond the scope of this chapter. Once again, readers should refer to the books cited under Further Reading to ensure that they are familiar with all the immune mechanisms (and jargon!) described below. However, a brief summary of some of the more pertinent aspects is well worth considering, beginning with the humoral immune response, which results in the production of antibodies.

Figure 6.1 shows a grossly simplified version of the mammalian humoral response to infection. Virus infection induces at least three classes of antibody: immunoglobulin G (IgG), IgM, and IgA. IgM is a large, multivalent molecule which is most effective at cross-linking large targets (e.g. bacterial cell walls or flagellae), but is probably less important in combating virus infections. In contrast, the production of IgA is very important for initial protection from virus infection. Secretory IgA is produced

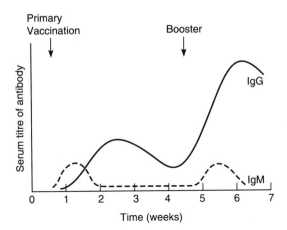

Primary
Vaccination

Booster

Serum titre of antibody

IgG

IgM

0   1   2   3   4   5   6   7

Time (weeks)

Figure 6.1 Simplified version of the kinetics of the mammalian humoral response to a 'typical' foreign virus (or other) antigen.

at mucosal surfaces and results in 'mucosal immunity,' an important factor in preventing infection from occurring. Induction of mucosal immunity depends to a large extent on the way in which antigens are presented to and recognized by the immune system. Similar antigens incorporated into different vaccine delivery systems (see Prevention and Therapy of Virus Infection) can lead to very different results in this respect, and mucosal immunity is such an important factor that similar vaccines may vary considerably in their efficacy. IgG is probably the most important class of antibody for direct neutralization of virus particles in serum and other body fluids (into which it diffuses).

Direct virus neutralization by antibodies results from a number of mechanisms, including conformational changes in the virus capsid caused by antibody binding, or blocking of the function of the virus target molecule (e.g. receptor binding) by steric hindrance. A secondary consequence of antibody binding is phagocytosis of antibody-coated ('opsonized') target molecules by mononuclear cells or polymorphonuclear lymphocytes. This process is mediated by the presence of the Fc receptor on the surface of these cells, but as has already been noted in Chapter 4, in some cases, opsonization of virus by the binding of non-neutralizing antibodies can result in enhanced virus uptake (e.g. with rabies virus) and possibly, in the case of HIV, uptake by macrophages.

Antibody binding also leads to the activation of the complement cascade, which assists in the neutralization of virus particles. The morphological alterations induced by the disruption of viruses by complement can sometimes be visualized directly by electron microscopy. The major impact of the humoral immune response is the eventual clearance of virus from the body, i.e. serum neutralization stops the spread of virus to uninfected cells and allows other defense mechanisms to mop up the infection.

In spite of all the above mechanisms, in overall terms, cell-mediated immunity is probably more important than humoral immunity in the control of virus infections. This is demonstrated by the following observations:

- Congenital defects in cell-mediated immunity tend to result in predisposition to viral (and parasitic) infections, rather than to bacterial infections.
- The functional defect in acquired immune deficiency syndrome (AIDS) is a reduction in the ratio of T-helper (CD4$^+$):T-suppressor (CD8+) cells from the normal value of about 1.2 to ≤ 0.2. AIDS patients commonly suffer many opportunistic virus infections (e.g. various herpesviruses, such as HSV, CMV, and EBV), which may have been present before the onset of AIDS but were previously suppressed by the intact immune system.

Cell-mediated immunity is effected via three main mechanisms: non-specific cell killing (mediated by 'natural killer' (NK) cells), specific cell killing (mediated by cytotoxic T-lymphocytes – CTL) and antibody-dependent cellular cytotoxicity (ADCC) (Figure 6.2). NK cells mediate cell lysis independently of conventional immunological specificity, i.e. clonal antigen recognition. They are not MHC-restricted and do not express Fc receptors and are therefore active without the requirement for sensitizing antibodies. The advantage of this is that they have broad specificity – they are the first line of defense against virus infection. NK cells are most active in the early stages of infection (i.e. the first few days) and their activity is stimulated by interferon-α (see below). NK cells are not directly induced by virus infection – they exist even in immunologically naive individuals and are 'revealed' in the presence of interferon-α. Their function is complementary to and is later taken over by CTL (below). The target for NK cells on the surface of infected cells is not known, but their action is inhibited by MHC class I antigens (which is present on all nucleated cells), allowing recognition of 'self' and preventing

Figure 6.2 Diagram illustrating the three main mechanisms by which cell-mediated immunity is effected.

total destruction of the body. It is known that some virus infections disturb normal cellular MHC class I expression and this is one possible mechanism of NK recognition of infected cells. The cell killing is achieved by the release of perforin (or 'cytolysin'), a protein found in NK cell cytoplasmic granules. Perforin is a peptide related to the complement protein C9 which, after release from the NK cells, polymerizes to form polyperforin. This forms transmembrane channels in the target cells, resulting in permeability of the target cell membrane and death.

Cytotoxic T-lymphocytes (CTL) are usually of CD8$^+$ (suppressor) phenotype. CTL are the major cell-mediated immune response to virus infections and are MHC-restricted, i.e. clones of cells recognize a specific antigen only in the context of MHC class II antigen plus the T-cell receptor/T3 complex on the surface of the CTL. (MHC class II antigens are expressed only on the surface of the antigen-presenting cells of the immune system – T-cells, B-cells, and macrophages.) This process requires 'help' (i.e. lymphokine production) from T-helper cells. The CTL themselves recognize foreign antigens through the TCR/T3 complex, which 'docks' with antigen presented by MHC class I on the surface of the target cell (Figure 6.3). However, the mechanism of cell killing by CTL is similar to that of NK cells, the production and release of perforin. The induction of a CTL response also results in the release of many different lymphokines from T-helper cells, some of

Figure 6.3 Cell surface proteins involved in immune recognition. Close contact between cells results in cell-to-cell signaling which regulate the immune response.

which result in clonal proliferation of antigen-specific CTL and others which have direct antiviral effects, for example, interferons (see below). The kinetics of the CTL response (which peak at about 7 days after infection) are somewhat slower than the NK response (e.g. 1–3 days) and, therefore, these two are complementary systems.

The induction of a CTL response is dependent on recognition of specific T-cell epitopes by the immune system. These are distinct from the B-cell epitopes recognized by the humoral arm of the immune system. T-cell epitopes are frequently more highly conserved (less variable) than B-cell epitopes, which are frequently able to mutate quickly to escape immune pressure. These are important considerations in the design of antiviral vaccines. The specificity of killing by CTL is not absolute. Although they are better 'behaved' than NK cells, diffusion of perforin and local lymphokine production frequently results in inflammation and bystander cell damage. This is a contributory cause of the pathology of many virus diseases (see Chapter 7), but the alternative is to allow virus replication to proceed unchecked!

ADCC is less well understood than either of the above two mechanisms. It can be mediated either by NK cells or by CTL. The mechanism of cell killing is the same as above, although complement may also be involved in ADCC. This mechanism is dependent on the recognition of antigen on the surface of the target cell by antibody on the surface of the effector cell. The antibody involved is usually IgG, which is bound to Fc receptors on the surface of the T-cell. ADCC therefore requires a pre-existing antibody response, and hence does not occur early during primary virus infections. The overall contribution of ADCC to the control of virus infections is not clear, although it is now believed that it plays a significant part in their control.

Another important factor in protection from virus infection is the production of interferons (IFN). By the 1950s, interference (i.e. the blocking of a virus infection by a competing virus) was a well-known phenomenon in virology. In some cases, the mechanism responsible is just that; for example, avian retroviruses are grouped into nine interference groups, A–I, based on their ability to infect various strains of chickens, pheasants, partridges, quail, etc., or cell lines derived from these species. In this case, the inability of particular viruses to infect the cells of some strains is due to the expression of the envelope glycoprotein of an endogenous provirus present in the cells, which sequesters the cellular receptor needed by the exogenous virus for infection. In other cases, the mechanism of viral interference is less clear.

In 1957, Alick Issacs and Jean Lindenmann were trying to study this phenomenon and performed the following experiment. Pieces of chick chorioallantoic membrane were exposed to UV-inactivated (non-infectious) influenza virus in tissue culture. The 'conditioned' medium from these experiments (which did not contain infectious virus) was found to inhibit the infection of fresh pieces of chick chorioallantoic membrane by (infectious) influenza virus in separate cultures. Their conclusion was that a soluble factor, which they called 'interferon,' was produced by cells as a result of virus infection and that this factor could prevent the infection of other cells. As a result of this provocative observation for many years interferon was the great hope for virology, and was thought to be directly equivalent to the use of antibiotics to treat bacterial infections.

The true situation has turned out to be far more complex than was first thought. Interferons do have antiviral properties, but by and large their effects are exerted indirectly via their major function as cellular regulatory proteins. Interferons are immensely potent; less than 50 molecules/cell shows evidence of antiviral activity. Hence, following Isaacs and Lindenmann's initial discovery, many fairly fruitless years were spent trying to purify minute amounts of naturally produced interferon. This situation changed with the development of molecular biology and the cloning and expression of interferon genes, which has led to rapid advances in our understanding over the last 15 years. There are three types of interferon – α, β, and γ.

- *Interferon-α*: There are at least 15 molecular species of interferon-α, all of which are closely related – some species differing by only one amino acid. They are synthesized predominantly by lymphocytes. The mature proteins contain 143 amino acids, with a minimum homology of 77% between the different types. All the genes encoding interferon-α are located on human chromosome 9 and gene duplication is thought to be responsible for this proliferation of genes.
- *Interferon-β*: There is a single gene for interferon-β, also on human chromosome 9. The mature protein contains 145 amino acids and, unlike interferon-α, is glycosylated, with approximately 30% homology to other interferons. It is synthesized predominantly by fibroblasts.
- *Interferon-γ*: There is a single gene for interferon-γ, on human chromosome 12. The mature protein contains 146 amino acids, is glycosylated, has very low sequence homology to other interferons. It is synthesized predominantly by lymphocytes.

Induction of interferon synthesis results from upregulation of transcription from the interferon gene promoters. There are three main mechanisms involved:

- *Virus infection*: This mechanism is thought to act by the inhibition of cellular protein synthesis which occurs during many virus infections, resulting in a reduction in the concentration of intracellular repressor proteins and hence in increased interferon gene transcription. In general, RNA viruses are potent inducers of interferon while DNA viruses are relatively poor inducers. However, there are exceptions to this rule (e.g. poxviruses are very potent inducers). The molecular events in the induction of interferon synthesis by virus infection are not clear. In some cases (e.g. influenza virus), UV-inactivated virus is a potent inducer. Therefore, virus replication is not necessarily required. Induction by viruses might involve perturbation of the normal cellular environment and/or production of small amounts of double-stranded RNA (see below).
- *Double-stranded (d/s) RNA*: All naturally occurring double-stranded RNAs (e.g. reovirus genomes) are potent inducers of interferon, as are synthetic molecules (e.g. poly I:C). Therefore, this process is independent of nucleotide sequence. Single-stranded RNA and double-stranded DNA are not inducers, hence the mechanism of induction is thought to depend on the secondary structure of the RNA.
- *Metabolic inhibitors*: Compounds which inhibit cellular transcription (e.g. actinomycin D) or translation (e.g. cycloheximide) result in induction of interferon. Tumor promoters such as TPA (tetradecanoyl phorbol acetate) or DMSO

(dimethyl sulfoxide) are also inducers – their mechanism of action remains unknown.

The effects of interferons are exerted via specific receptors on the surface of target cells. The main action is on cellular regulatory activities, and is rather complex. Interferon affects both cellular proliferation and immunomodulation. These effects result from the induction of transcription of a wide variety of cellular genes, including other lymphokines. The net result is complex regulation of a cell's ability to proliferate, differentiate, and communicate. This cell regulatory activity itself has indirect effects on virus replication (see below). All interferons have similar antiviral capacities, but IFN-γ is by far the most potent cellular regulator.

The effect of interferons on virus infections *in vivo* is extremely important, as has been shown by experiments in which animals experimentally infected with viruses and injected with anti-interferon antibodies experience much more severe infections than control animals infected with the same virus. This is because interferons protect cells from damage and death. However, they do not appear to play a major role in the clearance of virus infections – the other parts of the immune response are necessary for this. Interferon is a firebreak which inhibits virus replication in its earliest stages by several mechanisms. Two of these are understood in some detail (below), but there are a number of others (in some cases specific to certain viruses) which are less well understood.

Interferons induce transcription of a cellular gene encoding 2′,5′-oligo A synthetase (Figure 6.4). There are at least four molecular species of 2′,5′-oligo A, induced by different forms of interferon. This compound activates an RNA-digesting enzyme, RNAse L, which digests viral genomic RNAs, viral and cellular mRNAs, and cellular ribosomal RNAs. The end result of this mechanism is a reduction in protein synthesis (due to the degradation of mRNAs and rRNAs) and, therefore, the cell is protected from virus damage. The second method relies on the activation of a

Figure 6.4 Mechanism of induction of 2′,5′-oligo A synthetase by interferons.

68 kDa protein kinase (Figure 6.5). This enzyme phosphorylates a cellular factor, eIF2α, which is required by ribosomes for the initiation of translation. Therefore, the net result of this mechanism is also the inhibition of protein synthesis and this reinforces the 2′,5′-oligo A mechanism. A third, well-established mechanism depends on the M_x gene, a single-copy gene located on human chromosome 21, whose transcription is induced by interferons-α and -β but not -γ. The product of this gene inhibits the primary transcription of influenzavirus but not of other viruses. Its method of action is unknown. In addition to these three, there are many additional recorded effects of interferons. They inhibit the penetration and uncoating of SV40 and some other viruses, possibly by altering the composition/structure of the cell membrane; they inhibit the primary transcription of many virus genomes (e.g. SV40, HSV); and they inhibit cell transformation by retroviruses. None of the molecular mechanisms by which these effects are mediated have been fully explained.

In conclusion, it can be stated that interferons are a powerful weapon against virus infection, but that they act as a blunderbuss, rather than a 'magic bullet'! The severe side-effects (fever, nausea, malaise) which result from the powerful cell-regulatory action of interferons means that they will never be widely used for the treatment of trivial virus infections – they are not the cure for the common cold. However, as the cell-regulatory potential of interferons is becoming better understood, they are finding increasing use as a treatment for certain cancers (e.g. the use of interferon-α in the treatment of hairy cell leukemia). The long-term prospects for their use as antiviral compounds are less certain, except for possibly in life-threatening infections where there is no viable alternative therapy.

Interferons are an effective means of curbing the worst effects of virus infections. Part of their wide-ranging efficacy results from their generalized, non-specific effects (e.g. the inhibition of protein synthesis in virus-infected cells). This lack of specificity means that it is very difficult for viruses to evolve strategies to counteract their effects. Nevertheless, there are a few known instances where this has happened. The anti-interferon effect of adenovirus VA RNAs has already been described in

Figure 6.5 Mechanism of induction of p68 kinase by interferons. GEF: guanine nucleotide exchange factor.

Chapter 5. Vaccinia virus is also known to show resistance to the antiviral effects of interferons. One of the early genes of this virus appears to encode a protein which inhibits the action of the interferon-induced p68 protein kinase. In addition, vaccinia virus replication is also resistant to the 2,5-oligo A-induced RNAse L, although the mechanism by which this resistance is achieved is not known.

Virus–Host Interactions

At this point, it would be as well to state once again that viral pathogenesis is an abnormal situation of no value to the virus – the vast majority of virus infections are subclinical, i.e. asymptomatic. However, for pathogenic viruses, there are a number of critical stages in replication which determine the nature of the disease they produce. For all viruses, pathogenic or non-pathogenic, the first factor which influences the course of infection is the mechanism and site of entry into the body (Figure 6.6):

- *The skin*: Mammalian skin is a highly effective barrier against viruses. The outer layer (epidermis) consists of dead cells and therefore does not support virus replication. Very few viruses infect directly by this route unless there is prior injury such as minor trauma or puncture of the barrier, such as insect or animal bites, or subcutaneous injections. Some viruses which do use this route are herpes simplex virus and papillomaviruses, although these viruses probably still require some form of disruption of the skin such as small abrasions or eczema.
- *Mucosal membranes*: The mucosal membranes of the eye and genitourinary (GU) tract are much more favorable routes of access for viruses to the tissues of the

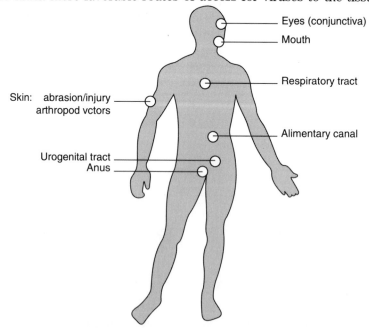

Figure 6.6 Schematic diagram to illustrate possible sites of virus entry into the body.

Table 6.1 ~~which~~ infect via mucosal surfaces

Virus	Site of infection
Adenoviruses	Conjunctiva
Picornaviruses: Enterovirus 70	Conjunctiva
Papillomaviruses	Genitourinary tract
Herpesviruses	Genitourinary tract
Retroviruses: HIV	Genitourinary tract
HTLV	Genitourinary tract

body. This is reflected by the number of viruses which can be sexually transmitted, and virus infections of the eye are also quite common (Table 6.1).

- *Alimentary canal*: Viruses may infect the alimentary canal via the mouth, oropharynx, gut, or rectum, although viruses which infect the gut via the oral route must survive passage through the stomach, an extremely hostile environment with a very low pH and high concentrations of digestive enzymes. Nevertheless, the gut is a highly valued prize for viruses – the intestinal epithelium is constantly replicating and there is a good deal of lymphoid tissue associated with the gut which provides many opportunities for virus replication. Moreover, the constant intake of food and fluids provides ample opportunity for viruses to infect these tissues (Table 6.2). To counteract this problem, the gut has many specific (e.g. secretory antibodies) and non-specific (e.g. stomach acids and bile salts) defense mechanisms.
- *Respiratory tract*: The respiratory tract is probably the most frequent site of virus infection. As with the gut, it is constantly in contact with external virus particles, taken in during respiration. As a result, the respiratory tract also has defenses aimed at virus infection – filtering of particulate matter in the sinuses, and cells and antibodies of the immune system present in the lower regions. Viruses which infect the respiratory tract usually come directly from the respiratory tract of others, since aerosol spread is very efficient: 'coughs and sneezes spread diseases' (Table 6.3).

The natural environment is a considerable barrier to virus infections. Most viruses are relatively sensitive to heat, drying, ultraviolet light (sunlight), etc., although a

Table 6.2 Viruses which infect via the alimentary canal

Virus	Site of infection
Herpesviruses	Mouth and oropharynx
Reoviruses	Intestinal tract
Calciviruses	Intestinal tract
Adenoviruses	Intestinal tract
Coronaviruses	Intestinal tract
Picornaviruses: Enteroviruses	Intestinal tract

Table 6.3 Viruses which infect via the respiratory tract

Virus		Localized infection
Orthomyxoviruses		Upper respiratory tract
Picornaviruses: Rhinoviruses		Upper respiratory tract
Coronaviruses		Upper respiratory tract
Adenoviruses		Upper respiratory tract
Paramyxoviruses:	Parainfluenza	Lower respiratory tract
	RSV	Lower respiratory tract

Virus	Systemic infection
Paramyxoviruses	Measles, mumps
Togaviruses	Rubella
Herpesviruses	Varicella
Poxviruses	Smallpox

few types are quite resistant to these factors. This is particularly important for viruses which are spread via contaminated water or foodstuffs – not only must they be able to survive in the environment until they are ingested by another host, but as most are spread by the fecal–oral route, they must also be able to pass through the stomach to infect the gut before being shed in the feces. One way of overcoming environmental stress is to take advantage of a secondary vector for transmission between the primary hosts (Figure 6.7). As with plant viruses (above), the virus may or may not replicate while in the vector. Viruses without a secondary vector must rely on continued host-to-host transmission, and have evolved various strategies to do this (Table 6.4):

- *Horizontal transmission*: The direct host-to-host transmission of viruses. This strategy relies on a high rate of infection to maintain the virus population.
- *Vertical transmission*: The transmission of the virus from one generation of hosts to the next. This may occur by infection of the fetus before, during, or shortly after birth (e.g. during breastfeeding). More rarely, it may involve direct transfer of the virus via the germ line itself, e.g. retroviruses. In contrast to horizontal transmission, this strategy relies on long-term persistence of the virus in the host rather than rapid propagation and dissemination of the virus.

Having gained entry to a potential host, the virus must initiate an infection by entering a susceptible cell (primary replication). This initial interaction frequently determines whether the infection will remain localized at the site of entry or spread to become a systemic infection (Table 6.5). In some cases, virus spread is controlled by infection of polarized epithelial cells and the preferential release of virus from either the apical (e.g. influenza virus – a localized infection in the upper respiratory tract) or basolateral (e.g. rhabdoviruses – a systemic infection) surface of the cells (Figure 6.8). Following primary replication at the site of infection, the next stage may be spread throughout the host. Apart from direct cell–cell contact, there are two main mechanisms for spread throughout the host:

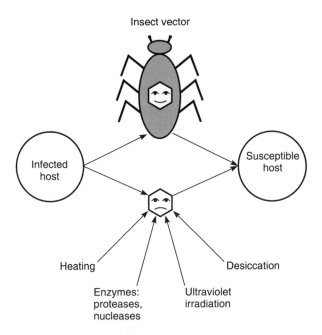

Figure 6.7 Transmission of viruses through the environment. Some viruses have adopted the use of vectors such as insects or other arthropods to avoid environmental stress.

Table 6.4 Virus transmission patterns

Pattern	Example
Horizontal transmission:	
Human–human (aerosol)	Influenza
Human–human (fecal–oral)	Rotaviruses
Animal–Human (direct)	Rabies
Animal–Human (vector)	Bunyaviruses
Vertical transmission:	
Placental–fetal	Rubella
Mother–child (birth)	HSV, HIV
Mother–child (breastfeeding)	HIV
Germ line	In mice, retroviruses; in humans ?

Table 6.5 Examples of localized and systemic infections

Virus	Site of primary replication	Secondary replication
Localized infections:		
Rhinoviruses	Upper respiratory tract	—
Rotaviruses	Intestinal epithelium	—
Papillomaviruses	Epidermis	—
Systemic infections:		
Enteroviruses	Intestinal epithelium	Lymphoid tissues, CNS
Herpesviruses	Oropharynx or GU tract	Lymphoid cells, CNS

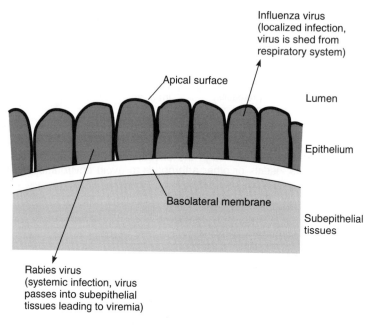

Figure 6.8 Virus infection of polarized epithelial cells.

- *Via the bloodstream*: Viruses may get into the bloodstream by direct inoculation, for example, by arthropod vectors, blood transfusion, or intravenous drug abuse (sharing of non-sterilized needles). The virus may travel free in the plasma (e.g. togaviruses, enteroviruses), or in association with red cells (orbiviruses), platelets (HSV), lymphocytes (EBV, CMV), or monocytes (lentiviruses). Primary viremia usually precedes and is necessary for the spread of virus to other parts of the body via the bloodstream, and is followed by a more generalized, higher titer secondary viremia as the virus reaches the other target tissues or replicates directly in blood cells.

- *Via the nervous system*: As above, spread of virus to the nervous system is usually preceded by primary viremia. In some cases, spread occurs directly by contact with neurons at the primary site of infection, in other cases it occurs via the bloodstream. Once in peripheral nerves, the virus can spread to the CNS by axonal transport along neurons. The classic example of this is herpes simplex virus (see Latent Infection, page 159). Viruses can cross synaptic junctions since these frequently contain virus receptors, allowing the virus to 'jump' from one cell to another.

The spread of the virus to various parts of the body is controlled to a large extent by its cell or tissue tropism. Tissue tropism is controlled partly by the route of infection, but largely by the interaction of a virus attachment protein with a specific receptor molecule on the surface of a cell (as discussed in Chapter 4), and has considerable effect on pathogenesis.

At this stage, following a significant degree of virus replication and the production of viral antigens, the host immune response comes into play. This has already been discussed (page 142) and obviously has a major impact on the outcome of an infection. To a large extent, the efficiency of the immune response determines the amount of secondary replication which occurs and hence, the spread to other parts of the body. If a virus can be prevented from reaching tissues where secondary replication can occur, generally no disease results, although there are some exceptions to this. The immune response also plays a large part in determining the amount of cell and tissue damage which occurs as a result of virus replication. As described above, the production of interferons is a major factor in preventing virus-induced tissue damage.

The immune system is not the only factor which controls cell death, the amount of which varies considerably for different viruses. Viruses may replicate widely throughout the body without any disease symptoms if they do not cause significant cell damage or death. Retroviruses do not generally cause cell death, being released from the cell by budding rather than by cell lysis, and cause persistent infections, even being passed vertically to the offspring if they infect the germ line. All vertebrate genomes, including humans, are littered with retrovirus genomes which have been with us for millions of years (Chapter 3). At the present time, these ancient virus genomes are not known to cause any disease in humans, although there are examples of tumors in rodents caused by them. Conversely, picornaviruses cause lysis and death of the cells in which they replicate, leading to fever and increased mucus secretion in the case of rhinoviruses, and paralysis or death (usually due to respiratory failure due to damage to the central nervous system resulting, in part, from viral replication in these cells) in the case of poliovirus.

The eventual outcome of any virus infection depends on a balance between two processes:

CLEARANCE ⇆ PERSISTENCE
(by the host organism) (by the virus)

Clearance is mediated by the immune system (as discussed previously). But the virus is a moving target which rapidly responds to pressure from the immune system by altering its antigenic composition (whenever possible). The classic example of this phenomenon is influenza virus, which displays two genetic mechanisms that allow the virus to alter its antigenic constitution:

- *Antigenic drift*: This involves the gradual accumulation of minor mutations (e.g. nucleotide substitutions) in the virus genome which result in subtly altered coding potential and therefore altered antigenicity, leading to decreased recognition by the immune system. This process occurs in all viruses all the time, but at greatly different rates; for example, it is much more frequent in RNA viruses than in DNA viruses. In response, the immune system constantly adapts by recognition of and response to novel antigenic structures – but it is always one step behind. In most cases, however, the immune system is eventually able to overwhelm the virus, resulting in clearance.
- *Antigenic shift*: In this process a sudden and dramatic change in the antigenicity of a virus occurs due to reassortment of the segmented virus genome with another genome of a different antigenic type (see Chapter 3). This results initially in the failure of the immune system to recognize a new antigenic type, giving the virus the upper hand (Figure 6.9).

The other side of the relationship that determines the eventual outcome of a virus infection is the ability of the virus to persist in the host. Long-term persistence of viruses results from two main mechanisms. The first is the regulation of lytic potential. The strategy followed here is to achieve the continued survival of a critical number of virus infected cells, i.e. sufficient to continue the infection without killing the host organism. For viruses which do not usually kill the cells in which they replicate, this is not usually a problem, hence these viruses tend naturally to cause persistent infections, e.g. retroviruses. For viruses that undergo lytic infection, e.g. herpesviruses, it is necessary to develop mechanisms which restrict virus gene expression, and consequently, cell damage (see below). The second aspect of persistence is the evasion of immune surveillance. Viruses have evolved many tricks to fool the immune system:

- Antigenic variation.
- Immune tolerance, causing a reduced response to an antigen, which may be due to genetic factors, pre-natal infection, or molecular mimicry by the virus of a host 'self' antigen, which is not recognized by immune system.
- Restricted gene expression.
- Downregulation of MHC class I gene expression, resulting in lack of recognition of infected cells, e.g. in adenovirus infections.
- Downregulation of expression of accessory molecules involved in immune recognition, e.g. LFA-3 and ICAM-1 (Figure 6.3) by EBV.
- Infection of immunocompromised sites within the body, e.g. HSV 'hides' in sensory ganglia in the nervous system.
- Direct infection of the cells of the immune system itself, e.g. herpesviruses, retroviruses (HIV) – often resulting in immunosuppression.

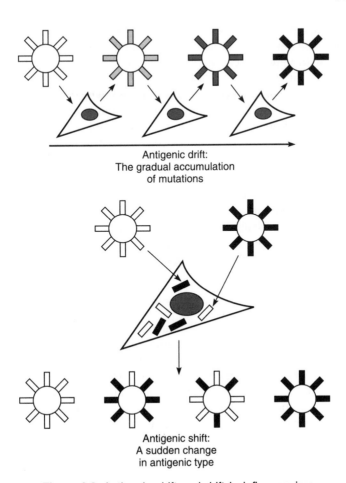

Antigenic drift:
The gradual accumulation
of mutations

Antigenic shift:
A sudden change
in antigenic type

Figure 6.9 Antigenic shift and drift in influenzavirus.

The Course of Virus Infections

Patterns of virus infection can be divided into a number of different types.

Abortive infection

This occurs when a virus infects a cell (or host), but cannot complete the full replication cycle. Therefore, this is a non-productive infection. However, the outcome of such infections is not necessarily insignificant; for example, SV40 infection of non-permissive rodent cells sometimes results in transformation of the cells (see Chapter 7).

Acute infection

This pattern is familiar from many common virus infections (e.g. 'colds'). In these relatively brief infections, the virus is usually eliminated completely by the immune system, but occasionally can have lasting effects; for example, herpes zoster infection causes chicken pox (an acute, self-limited disease), but also shingles (a chronic disease with neurological involvement caused by reactivation of the latent virus – see Latent Infection). Typically in acute infections, most viral replication occurs before the onset of any symptoms (e.g. fever), which are the result not only of viral replication but of the activation of the immune system. Therefore, acute infections present a serious problem for the epidemiologist and are the pattern most frequently associated with epidemics (e.g. influenza, measles, etc.).

Chronic infection

These are the converse of acute infections, i.e. prolonged and stubborn. To cause this type of infection, the virus must persist in the host for a significant period. To the clinician, there is no clear distinction between chronic, persistent, and latent infections and the terms are often used interchangeably. They are listed separately here because to virologists, there are significant differences in the events which occur during these infections.

Persistent infection

These infections result from a delicate balance between the virus and the host organism, in which ongoing virus replication occurs, but the virus adjusts its replication and pathogenicity so as to avoid killing host. They differ from chronic infections in that whereas in chronic infections the virus is usually eventually cleared by the host (unless the infection proves fatal), in persistent infections, the virus may continue to be present and to replicate in the host for its entire lifetime.

The best studied example of such a system is lymphocytic choriomeningitis virus (LCMV – an arenavirus) infection in mice (Figure 6.10). Mice can be experimentally

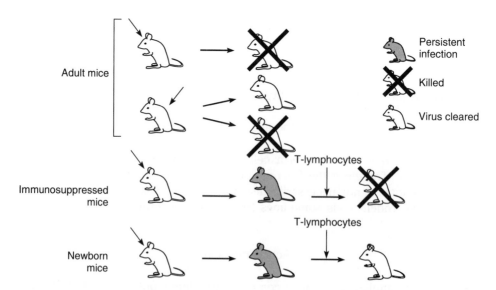

Figure 6.10 Persistent infection of mice by lymphocytic choriomeningitis virus (LCMV).

infected with this virus either at a peripheral site (e.g. a footpad or the tail) or by direct inoculation into the brain. Adult mice infected in the latter way are killed by the virus, but if infected by a peripheral route, there are two possible outcomes to the infection. Some mice die after peripheral infection, whereas others survive, having cleared the virus from the body completely. It is not clear what factors determine the survival or death of LCMV-infected mice, but other evidence shows that this is related to the immune response to the virus. In immunosuppressed adult mice which are infected via the CNS route, a persistent infection is established in which the virus is not cleared (due to the non-functional immune system), but remarkably, these mice are not killed by the virus. If, however, syngeneic LCMV-specific T-lymphocytes (i.e. of the same MHC type) are injected into these persistently infected mice, the animals develop the full pathogenic symptoms of LCMV infection and are killed. Newborn mice, whose immune systems are immature, infected via the CNS route also develop a persistent infection, but in this case, if subsequently injected with syngeneic LCMV-specific T-lymphocytes, they clear the virus and survive the infection. The mechanisms which control these events are not completely understood, but it is evident that there is a delicate balance between the virus and the host animal and that the immune response to the virus is partly responsible for the pathology of the disease and the death of the animals.

Not infrequently, persistent infections may result from the production of defective-interfering (DI) particles (see Chapter 3). Such particles contain a partial deletion of the virus genome and are replication defective, but are maintained and may even tend to accumulate during infections as they can replicate in the presence of replication-competent helper virus. The production of DI particles is a common consequence of virus infection of animals, particularly by RNA viruses, but also occurs with DNA viruses and plant viruses, and can be mimicked *in vitro* by continuous high titer passage of virus. Although not able to replicate themselves independently, DI particles are not necessarily genetically inert, and may alter the course of an infection by recombination with the genome of a replication-competent virus. The presence of DI particles can profoundly influence the course and the outcome of a virus infection. In some cases, they appear to moderate pathogenesis, whereas in others they potentiate it, making the symptoms of the disease much more severe. Moreover, as DI particles effectively cause restricted gene expression (since they are genetically deleted), they may also result in a persistent infection by a virus which normally causes an acute infection and is rapidly cleared from the body.

Latent infection

This is the ultimate infection! In latent infection the virus is able to downregulate its gene expression and establish a truly latent state, i.e. with strictly limited gene expression and without ongoing virus replication. Latent virus infections typically persist for the entire life of the host. An example of such an infection in man is herpes simplex virus (HSV). Infection via the skin or mucosa results in localized primary replication. During this phase, the virus infects peripheral neurons and travels via axonal transport mechanisms further into the nervous system. There, the virus 'hides' in dorsal root ganglia, such as the trigeminal ganglion, establishing a truly latent infection. The nervous system is an immunologically privileged site and is not patrolled by the immune system in the same way as the rest of the body, but nevertheleses, the major factor in latency is the ability of the virus to restrict its gene expression. This eliminates the possibility of recognition of infected cells by the immune system. Restricted gene expression is achieved by tight regulation of α-gene expression, which is an essential control point in herpesvirus replication (see Chapter 5). Recently, there has been much interest in mRNAs which are made by HSV during latent infection, known as latency-associated transcripts or LATs, raising the possibility of parallels between HSV latency and lysogeny in phage λ. However, the importance of the LATs in controlling latency has more recently been questioned, and the mechanism by which they might act is not clear at present.

When reactivated by some provocative stimulus, HSV travels down the sensory nerves to cause peripheral manifestations such as cold sores or genital ulcers. It is not altogether clear what constitutes a provocative stimulus, but there are many possible alternatives, including psychological as well as physical factors. Periodic reactivation establishes the pattern of infection, with sporadic, sometimes very painful reappearance of disease symptoms for the rest of the host's life. Even worse than this, any cause of immunosuppression later in life can cause the latent infection to flare up (which indicates that the immune system normally has a role in helping to suppress these latent infections), resulting in a very severe, systemic, and often life-threatening infection.

In a somewhat similar way to herpesviruses, infection by retroviruses may result in a latent infection. Integration of the provirus into the host genome certainly results in the persistence of the virus for the lifetime of the host organism and may lead to an episodic pattern of disease. In some ways, acquired immunodeficiency syndrome (AIDS), which results from HIV infection, shows aspects of this pattern of infection. The pathogenesis of AIDS is discussed in detail in Chapter 7.

Prevention and Therapy of Virus Infection

There are two aspects of the response to the threat of virus diseases: first, prevention of infection, and secondly, treatment of the disease. The former strategy relies on two approaches: public and personal hygiene, which perhaps plays the major role in preventing virus infection (e.g. provision of clean drinking water and disposal of sewage, and good medical practice, such as the sterilization of surgical instruments), and vaccination, which makes use of the immune system to combat virus infections. Most of the damage to cells during virus infections occurs very early, often before the clinical symptoms of disease appear. This makes the treatment of virus infection very difficult, and therefore, in addition to being cheaper, prevention of virus infection is undoubtedly better than cure. Although prevention of infection is much the preferred option, post-exposure vaccines can be of great value in modifying the course of some virus infections (e.g. rabies).

To design effective vaccines, it is important to understand both the immune response to virus infection (above) and the stages of virus replication that are appropriate targets for immune intervention. To be effective, vaccines must stimulate as many of the body's defense mechanisms as possible. In practice, this usually means trying to mimic the disease, without of course causing pathogenesis; for example, the use of nasally administered influenza vaccines and orally administered poliovirus vaccines. To be effective, it is not necessary to get 100% uptake of vaccine. 'Herd immunity' results from the break in transmission of a virus which occurs when a sufficiently high proportion of a population has been vaccinated. This strategy is most effective where there is no alternative host for the virus, e.g. measles, and in practice is the situation that usually occurs since it is impossible to achieve 100% coverage with any vaccine. However, this is a risky business; if protection of the population falls below a critical level, epidemics can easily occur. There are three basic types of vaccine: subunit vaccines, inactivated vaccines, and live virus vaccines.

Subunit vaccines

These are the newest type of vaccine and consist of only a part of the virus capsid, sufficient to induce a protective immune response but not enough to allow any danger of infection. In general terms, they are completely safe, except for very rare cases in which adverse immune reactions may occur. Unfortunately, at the present time, they are also the least effective and most expensive type of vaccines. The major technical problems associated with subunit vaccines are their relatively poor antigenicity and the need for new delivery systems, such as improved carriers and **adjuvants**. There are several categories of such vaccines:

- *Synthetic vaccines*, such as short, chemically synthesized peptides. The major disadvantage with these molecules is that they are not usually very effective immunogens and are very costly to produce. However, since they can be made to order for any desired sequence, they have great potential for the future. None are currently in use.
- *Recombinant vaccines*, produced by genetic engineering. Such vaccines have been already produced and are better than the above type since they tend to give rise to a more effective immune response. Some practical success has already been achieved with this type of vaccine. For example, vaccination against hepatitis B virus (HBV) used to rely on the use of 'Australian antigen' (HBsAg) obtained from the serum of chronic HBV carriers. This was a very risky practice indeed (since HBV carriers are often also infected with HIV). A completely safe recombinant HBV vaccine produced in yeast is now widely used.
- *Virus vectors*, i.e. recombinant virus genomes genetically manipulated to express protective antigens from (unrelated) pathogenic viruses. The idea here is to utilize the genome of a well-understood, attenuated virus to express and present antigens to the immune system. Many different viruses offer possibilities of this type of approach, but the most highly developed system so far is based on the vaccinia virus genome. This virus has been used to vaccinate millions of people worldwide in the campaign to eradicate smallpox (see below), and is generally a very safe and effective vehicle for antigen delivery. Such vaccines are difficult to produce and none is clearly successful yet, although many different trials are currently underway, e.g. vaccinia–rabies recombinants which have been used to eradicate rabies in European fox populations.

Inactivated vaccines

These are produced by exposing the virus to a denaturing agent under precisely controlled conditions. The objective is to cause loss of virus infectivity without loss of antigenicity. Obviously, this is treading a thin line. However, inactivated vaccines have certain advantages, such as generally being effective immunogens (if properly inactivated), being relatively stable, and carrying little or no risk of vaccine-associated virus infection (if properly inactivated, but accidents can and do occur). The disadvantage of these vaccines is that it is not possible to produced inactivated vaccines for all viruses, since denaturation of virus proteins may lead to loss of antigenicity, e.g. measles virus. Although relatively effective, 'killed' vaccines are sometimes not as effective at *preventing* infection as 'live' virus vaccines (see below), often because they fail to stimulate protective mucosal immunity to the same extent. A more recent concern is that these vaccines contain viral nucleic acids, which may themselves be a source of infection, either of their own accord (e.g. (+)sense RNA virus genomes) or after recombination with other viruses.

Live (attenuated) virus vaccines

This strategy relies on the use of viruses with reduced pathogenicity to stimulate an immune response without causing disease. The vaccine strain may be a naturally occurring virus (e.g. the use of cowpox virus by Edward Jenner to vaccinate against smallpox) or artificially **attenuated** *in vitro* (e.g. the oral poliomyelitis vaccines produced by Albert Sabin). The advantages of attenuated vaccines is that they are good immunogens and induce long-lived, appropriate immunity. Set against this are their many disadvantages. They are often biochemically and genetically unstable and may either lose infectivity (when they become worthless) or revert to virulence unexpectedly. In spite of intensive study, it is not possible to produce an attenuated vaccine to order, and there seems to be no general mechanism by which different viruses can be reliably and safely attenuated. Contamination of the vaccine stock with other, possibly pathogenic viruses is also possible – this was the way in which SV40 was first discovered in oral poliovirus vaccine in 1960. Inappropriate use of live virus vaccines, for example, in immunocompromised hosts or during pregnancy, may lead to vaccine-associated disease, whereas the same vaccine given to a healthy individual may be perfectly safe.

In spite of these difficulties, vaccination against virus infection has been one of the great triumphs of medicine during the twentieth century, and most of the success stories result from the use of live attenuated vaccines; for example, the use of vaccinia virus against smallpox, the first virus disease to be completely eradicated from the world. The next targets for eradication are poliomyelitis and measles, currently targeted by the World Health Organization. In the future, it is certain that these advances will be continued by the development and application of vaccines produced by recombinant DNA technology. Moreover, viruses are being developed as gene delivery systems for the treatment of inherited as well as acquired diseases. The first human trials using recombinant retrovirus genomes to treat children with immunodeficiency resulting from a lack of the enzyme adenosine deaminase (ADA) are underway. It is important to note that such experiments are aimed at augmenting the somatic cells of patients to alleviate the symptoms of the disease, and not at manipulating the human germ line, which is quite a different issue. There can be no question that carefully applied, this new application of virology will change the treatment of inherited diseases during the next century.

The alternative to vaccination is to attempt to treat virus infections using drugs which block virus replication (Table 6.6). Historically, discovery of antiviral drugs has been largely fortuitous. Spurred on by successes in the treatment of bacterial infections with antibiotics, drug companies launched huge blind-screening programs to identify chemical compounds with antiviral activity – with relatively little success. The key to the success of any antiviral drug lies in its specificity. Almost any stage of virus replication can be a target for a drug, but the drug must be more toxic to the virus than the host. This is measured by the chemotherapeutic index, given by:

$$\frac{\text{Dose of drug which inhibits virus replication}}{\text{Dose of drug which is toxic to host}}$$

The smaller the value of the chemotherapeutic index, the better. In practice, a difference of several orders of magnitude between the two toxicity values is usually required to produce a safe and clinically useful drug. Modern technology, including molecular biology and computer-aided design of chemical compounds, allows the deliberate design of drugs, but to do this, it is necessary to 'know your enemy,' i.e. to understand the key steps in virus replication which might be inhibited. Any of the stages of viral replication can be a target for antiviral intervention. The only requirements are:

(1) that the process targeted be essential for replication; and
(2) that the drug is active against the virus but has "acceptable toxicity" to the host organism.

Table 6.6 Antiviral drugs

Drug	Viruses	Chemical type	Target
Vidarabine	Herpesviruses	Nucleoside analog	Virus polymerase
Acyclovir	Herpes simplex	Nucleoside analog	Virus polymerase (needs virus thymidine kinase for activation)
Ganciclovir	Cytomegalovirus	Nucleoside analog	Virus polymerase (needs virus kinase for activation)
Zidovudine (AZT)	HIV	Nucleoside analog	Reverse transcriptase
Ribavirin	HSV, measles, mumps, lassa fever	Triazole carboxamide	Virus replicase/transcriptase (broad spectrum)
Amantadine/ rimantadine	Influenza A strains	Tricyclic amines	Matrix protein/hemagglutinin

What degree of toxicity is 'acceptable' clearly varies considerably, for example, between a cure for the common cold, which might be sold over the counter and taken by millions of people, and a drug used to treat fatal virus infections such as AIDS.

The attachment phase of replication can be inhibited in two ways: by agents which mimic the virus-attachment protein (VAP) and bind to the cellular receptor, or by agents which mimic the receptor and bind to the VAP. Synthetic peptides are the most logical class of compound to use for this purpose. While this is a promising line of research, there are considerable problems with clinical use of these substances, mostly the high cost of synthetic peptides and the poor pharmacokinetic properties of many of these synthetic molecules.

It is difficult to specifically target the penetration/uncoating stages of virus replication as relatively little is known about them. Uncoating in particular is largely mediated by cellular enzymes and is therefore a poor target for intervention, although like penetration, it is often influenced by one or more virus proteins. Amantadine and rimantadine are two drugs which are active against influenza A viruses. The action of these closely related agents is complex and incompletely understood, but they are believed to block cellular membrane ion channels. The target for both drugs is the matrix protein (M_2), but resistance to the drug maps to the hemagglutinin (HA) gene. This biphasic action results from the inability of drug-treated cells to lower the pH of the endosomal compartment (a function normally controlled by the M_2 gene product), a process which is essential to induce conformational changes in the HA protein to permit membrane fusion (see Chapter 4).

Many viruses have evolved their own specific enzymes to preferentially replicate virus nucleic acids at the expense of cellular molecules. There is often sufficient specificity in virus polymerases to provide a target for an antiviral agent, and this method has produced the majority of the specific antiviral drugs currently in use (Table 6.6). The majority of these drugs function as polymerase substrates (i.e. nucleoside/nucleotide) analogs, and their toxicity varies considerably from some which are well tolerated (e.g. acyclovir) to others which are quite toxic (e.g. AZT). There is a problem with the pharmacokinetics of these nucleoside analogs – their typical serum half-life is 1–4 h. Nucleoside analogs are, in fact, pro-drugs, since they need to be phosphorylated before becoming effective. This is the key to their selectivity:

- Acyclovir is phosphorylated by HSV thymidine kinase 200 times more efficiently than by cellular enzymes.
- Gancyclovir is 10 times more effective against CMV than acyclovir, but must be phosphorylated by a kinase encoded by CMV gene *UL97* before it becomes pharmaceutically active.

Virus gene expression is less amenable to chemical intervention than genome replication, since viruses are much more dependent on the cellular machinery for transcription, mRNA splicing, cytoplasmic export, and translation than for replication. So far, no clinically useful drugs which discriminate between viral and cellular gene expression have been developed. As with penetration and uncoating, for the majority of viruses, the processes of assembly, maturation, and release are poorly understood, and, therefore, have not as yet become targets for antiviral intervention.

Summary

Virus infection is a complex, multi-stage interaction between the virus and the host organism. The course and eventual outcome of any infection is the result of a balance between host and virus processes. Host factors involved include exposure to different routes of virus transmission and the control of virus replication by the immune response. Virus processes include the initial infection of the host, spread throughout the host and the regulation of gene expression to evade the immune response. Medical intervention against virus infections includes the use of vaccines to stimulate the immune response and drugs to inhibit virus replication. Molecular biology is stimulating the production of a new generation of antiviral drugs and vaccines.

Further Reading

Mims, C.A. (1987). *The Pathogenesis of Infectious Disease* (3rd edition). Academic Press, London.
Roitt, I. (1988). *Essential Immunology* (6th edition). Blackwell Scientific, Oxford.

Bibliography

Brochier, B. *et al.* (1991). Large scale eradication of rabies using recombinant vaccinia–rabies vaccine. *Nature* **354**: 520–2.
Brown, F. (Ed.) (1990). Modern approaches to vaccines. *Seminars in Virology*, **1**.
Garcia-Blanco, M.A. and Cullen, B.R. (1991). Molecular basis of latency in pathogenic human viruses. *Science* **254**: 815–20.
Ho, D.Y. (1992). Herpes simplex virus latency – molecular aspects. *Prog. Med. Virol.* **39**: 76–115.
Miller, A.D. (1992). Human gene therapy comes of age. *Nature* **357**: 455–60.
Oldstone, M.A. (1991). Molecular anatomy of viral persistence. *J. Virol.* **65**: 6381–6.

Chapter 7

Pathogenesis

Pathogenesis, which is the capacity of one organism to cause disease in another, is a complex and variable phenomenon. For one thing, it is a rather difficult character to define. At the simplest level there is the question, what is disease? An all-embracing definition would be that it is a departure from the normal physiological parameters of an organism. This could range from a transient and very minor condition such as a slightly elevated temperature or rather subjective feelings of lethargy, to chronic pathologic conditions which eventually result in death. Any of these conditions may result from a tremendous number of internal or external sources. However, there is rarely one single factor which 'causes' a disease. Most disease states are multi-factorial at one level or another.

Considering only viral diseases, there are two components involved: the direct effects of virus replication and the effects of bodily responses to the infection. The course of any virus infection is determined by a delicate and dynamic balance between the host and the virus, as described in Chapter 6. The extent and severity of viral pathogenesis is determined similarly. In some virus infections, most of the pathologic symptoms observed are attributable not to viral replication, but to the side-effects of the immune response. Inflammation, fever, headaches, and skin rashes are often not caused by viruses themselves, but by the cells of the immune system due to the release of potent chemicals such as interferons and interleukins. In the most extreme cases, it is possible that none of the pathologic effects of certain diseases are caused directly by the virus, excepting that its presence stimulates the activation of the immune system.

Viral pathogenesis is an abnormal and fairly rare situation. The vast majority of virus infections are silent and do not result in outward signs of disease. It is sometimes said that viruses would disappear if they killed their hosts. This is not necessarily true. It is perfectly possible to envisage viruses with a hit-and-run strategy, moving quickly from one dying host to the next and relying on continuing circulation for their survival. Nevertheless, there is a clear tendency for viruses not to injure their hosts if possible. A good example of this is the rabies virus. The symptoms of human rabies virus infections are truly dreadful, but thankfully very rare. However, in its normal hosts (e.g. foxes), rabies virus infection produces a much milder disease which does not usually kill the fox. Man is an unnatural, dead-end host for this virus and the severity of human rabies is extreme as the condition is rare. Ideally, a virus would not even provoke an immune response from its host, or at least be able to hide to avoid the effects. Herpesviruses and some retroviruses have evolved complex lifestyles which enable them to get close to this objective. Of course, fatal infections such as rabies and AIDS always grab the headlines. Much less effort has been devoted to isolating and studying the myriad viruses which have not (yet) caused well-defined diseases in man, his domestic animals, or economically valuable crop plants.

In the past few decades, molecular genetic analysis has contributed enormously to our understanding of viral pathogenesis. Nucleotide sequencing and site-directed mutagenesis have been used to explore molecular determinants of virulence in many different viruses. Specific sequences and structures which are found only in disease-causing strains of virus and not in closely related **attenuated** or **avirulent** strains have been identified. Sequence analysis has also led to the identification of T-cell and B-cell epitopes on virus proteins responsible for their recognition by the immune system. Unfortunately, these advances do not automatically lead to an understanding of the mechanisms responsible for pathogenicity.

Unlike the rest of this book, this chapter is specifically about viruses which cause disease in animals. It does not discuss bacteriophages, for which the terms 'pathogenesis' and 'disease' are hardly appropriate, nor viruses which cause disease in plants, which is a highly specialist field. Three major aspects of viral pathogenesis are considered: direct cell damage resulting from virus replication, damage resulting from immune activation or suppression, and cell transformation caused by viruses.

Mechanisms of Cellular Injury

Virus infection results in a number of changes which are detectable by visual or biochemical examination of infected cells. These changes result from the production of viral proteins and nucleic acids, but also from alterations to the biosynthetic capabilities of cells. Intracellular parasitism by viruses sequesters cellular apparatus such as ribosomes and raw materials which would normally be devoted to synthesizing molecules required by the cell. Eukaryotic cells must carry out constant macromolecular synthesis, whether they are growing and dividing or in a state of quiescence. A growing cell clearly needs to manufacture more proteins, more nucleic acids and more of all of its myriad components in order to increase its size before dividing. However, there is a much more fundamental requirement for such activity. The function of all cells is regulated by controlled expression of their genetic information and the subsequent degradation of the molecules produced. Such control thus relies on a delicate and dynamic balance between synthesis and decay which determines the intracellular levels of all the important molecules in the cell. This is particularly true of the control of the cell cycle, which determines the behavior of dividing cells (see Cellular Transformation by Viruses, page 177). Thus any interruption of the synthesis of these vital messages, or the addition of any virus-encoded protein which interferes with this process, has severe consequences for the cell. Even non-dividing cells must constantly synthesize a variety of macromolecules which require constant replacement. This activity is known as 'housekeeping' and all cells contain many thousands of these housekeeping genes.

In general terms, a number of common phenotypic changes can be recognized in virus-infected cells. These changes are often referred to as the **cytopathic effects** (c.p.e.) of a virus, and include:

- *Altered shape*: Adherent cells which are normally attached to other cells (*in vivo*) or an artificial substrate (*in vitro*) may assume a rounded shape different from their normal flattened appearance. The extended 'processes' (extensions of the cell surface resembling tendrils) involved in attachment or mobility are withdrawn into the cell.
- *Detachment from the substrate*: For adherent cells, this is the stage of cell damage which follows that above. Both of these effects are caused by partial degradation or disruption of the cytoskeleton which is normally responsible for maintaining the shape of the cell.
- *Lysis*: This is the most extreme case, where the entire cell breaks down. Membrane integrity is lost and the cell may swell due to the absorption of extracellular fluid and finally, break open. This is an extreme case of cell damage and it is important to realize that not all viruses induce this effect, although they may cause other cytopathic effects. Lysis is beneficial to a virus in that it provides an obvious method of releasing new virus particles from an infected cell. However, there are alternative ways of achieving this, such as release by budding (see Chapter 4).
- *Membrane fusion*: The membranes of adjacent cells fuse together, resulting in a mass of cytoplasm containing more than one nucleus, known as a **syncytium**, or, depending on the number of cells fused together, a giant cell. Fused cells are short lived and subsequently lyse – apart from direct effects of the virus, they cannot tolerate more than one non-synchronized nucleus per cell.
- *Inclusion bodies*: These are areas of the cell where virus components have accumulated. They are frequently sites of virus assembly, and some cellular inclusions consist of crystalline arrays of virus particles. It is not clear how these structures damage the cell, but they are frequently associated with viruses which cause cell lysis, such as herpesviruses and rabies virus.

In some cases, a great deal of detail is known about the molecular mechanisms of cell injury. A number of viruses which cause cell lysis exhibit a phenomenon known as **shutoff** early in infection. Shutoff is the sudden and dramatic cessation of most host cell macromolecular synthesis. In poliovirus-infected cells, this is the result of production of the virus 2A protein. This molecule is a protease which cleaves the p220 component of eIF–4F, a complex of proteins required for cap-dependent translation of messenger RNAs by ribosomes. Since poliovirus RNA does not have a 5′ methylated cap but is modifed by the addition of the VPg protein, virus RNA continues to be translated. In poliovirus-infected cells, the dissociation of mRNAs and polyribosomes from the cytoskeleton can be observed and this is the reason for the inability of the cell to translate its own messages. A few hours after translation ceases, lysis of the cell occurs.

In other cases, shutoff results from a different molecular mechanism. For many viruses, the sequence of events which occurs is not known. In the case of adenoviruses, the penton protein (part of the virus capsid) has a toxic effect on cells. Although its precise action on cells is now known, addition of purified penton protein to cultured cells results in their rapid death. Toxin production by pathogenic bacteria is a common phenomenon, but this is the only well-established case of a virus-encoded molecule with a toxin-like action. However, some of the normal contents of cells released on lysis may have toxic effects on other cells, and antigens which are not recognized as 'self' by the body (e.g. nuclear proteins) may result in immune activation and inflammation.

Membrane fusion is the result of virus-encoded proteins required for infection of cells (see Chapter 4), typically, the glycoproteins of enveloped viruses. One of the best known examples of such a protein comes from Sendai virus (a paramyxovirus), which has been used to induce cell fusion during the production of monoclonal antibodies (Chapter 1). Herpes simplex virus (HSV) encodes at least eight glycoproteins (Table 7.1). As can be seen, a number of these proteins are involved in fusion of the virus envelope with the cell membrane and also in cell penetration. Production of fused syncytia is a common feature of HSV infection.

Another virus which causes cell fusion is human immunodeficiency virus (HIV). Infection of CD4$^+$ cells with some but not all isolates of HIV causes cell–cell fusion and the production of syncytia or giant cells. The protein responsible for this is the transmembrane envelope glycoprotein of the virus and the domain near the amino terminus responsible for this fusogenic activity has been identified by molecular genetic analysis (Figure 7.1). Since HIV infects CD4$^+$ cells and it is the reduction in the number of these crucial cells of the immune system which is the most obvious defect in AIDS, it was initially believed that direct killing of these cells by the virus was the basis for the pathogenesis of AIDS. It has subsequently become clear that this is not the case. Although direct cell fusion and killing by the virus undoubtedly does occur in patients, the extent of this activity appears to be insufficient to account for the reduction in CD4$^+$ cell numbers in AIDS. It is now believed that the pathogenesis of AIDS is considerably more complex (see Viruses and Immunodeficiency, page 170). Many animal retroviruses also cause cell killing and in most cases, it appears that the envelope protein of the virus is required, although there may be more than one mechanism involved.

Table 7.1 The glycoproteins of herpes simplex virus

Protein	Functions
gpB	Fusion of virus envelope with cell membrane
gpC	Fusion of virus envelope with cell membrane
gpD	Major neutralization antigen; also involved in membrane fusion
gpE	Virion Fc receptor
gpG	Type-specific antigen
gpH	Involved in cell penetration (?)
gpI	Virion Fc receptor
gpJ	Predicted from nucleotide sequence, derived from gpH

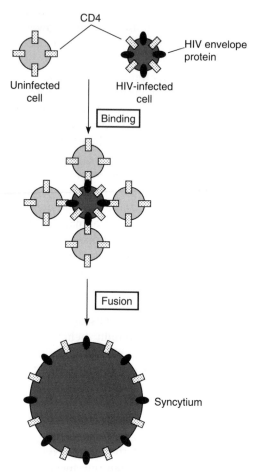

Figure 7.1 Mechanism of HIV-induced cell fusion. The virus envelope glycoprotein, whose role on virus particles is in receptor binding and membrane fusion, is expressed on the surface of infected cells. Uninfected CD4$^+$ cells coming into contact with these infected cells are fused together to form a **syncytium**.

Viruses and Immunodeficiency

At least two groups of viruses, herpesviruses and retroviruses, directly infect the cells of the immune system. This has important consequences for the outcome of the infection and for the immune system of the host. Herpes simplex virus (HSV) establishes a systemic infection, spreading via the bloodstream in association with platelets, but does not show a particular tropism for cells of the immune system. However, herpes saimirii and Marek's disease virus are herpesviruses which cause lymphoproliferative diseases (but not clonal tumors) in monkeys and chickens respectively. Human herpesvirus-6 (HHV-6) was first identified in lymphocytes from patients with lymphoreticular disease in 1986. HHV-6 infection is now known to be common in apparently healthy people worldwide. The main target cells for this virus are peripheral blood mononuclear cells (both T- and B-cells). Infection is associated with a skin rash which is particularly evident in children but there are no other apparent pathogenic symptoms.

Epstein–Barr virus (EBV) infection of B-cells leads to their immortalization and proliferation, resulting in 'glandular fever' or mononucleosis, a debilitating but benign condition. EBV was first identified in a lymphoblastoid cell line derived from Burkitt's lymphoma and, in rare instances, EBV infection may lead to the formation of a malignant tumor (see Cellular Transformation by Viruses, page 177). While some herpesviruses such as HSV are notably cytopathic, most of the lymphotropic herpesviruses do not cause a significant degree of cellular injury. However, infection of the delicate cells of the immune system may perturb their normal function. Since the immune system is internally regulated by complex networks of interlinking signals, relatively small changes in cellular function can result in its collapse. Alteration of the normal pattern of production of cytokines could have profound effects on immune function. The *trans*-regulatory proteins involved in the control of herpesvirus gene expression may also affect the transcription of cellular genes. Therefore, the effects of herpesviruses on immune cells are more complex than just cell killing.

Many retroviruses cause pathogenic conditions including paralysis, arthritis, anemia, and malignant cellular transformation. A significant number of retroviruses infect the cells of the immune system. Although these infections may lead to a diverse array of diseases and hematopoetic abnormalities such as anemia and lymphoproliferation, the most commonly recognized consequence of retrovirus infection is the formation of lymphoid tumors (see Cellular Transformation by Viruses). However, some degree of immunodeficiency, ranging from very mild to quite severe, is a common consequence of the interference with the immune system resulting from the presence of a lymphoid or myeloid tumor.

In the last decade, however, the most prominent aspect of virus-induced immunodeficiency has been acquired immunodeficiency syndrome (AIDS), which is a consequence of infection with human immunodeficiency virus (HIV), a member of the lentivirus group of the retroviruses. There are a number of similar lentiviruses which cause immunodeficiency diseases in primates, cats, and mice as well as man. Unlike infection by other types of retrovirus, HIV infection does not directly result in the formation of tumors. Some tumors such as B-cell lymphomas are sometimes seen in AIDS patients but are probably a consequence of the lack of immune surveillance which is responsible for the destruction of tumors in healthy individuals. The clinical course of AIDS is long and very variable. A great number of different abnormalities of the immune system are seen in AIDS (Table 7.2). As a result of the biology of lentivirus infections, the pathogenesis of AIDS is highly complex. Some of the relevant factors to consider are:

- *Latency*: Lentiviruses do not show true latency (unlike herpesviruses or phage λ), but they do have the capacity to control the expression of their genomes by means of virus-encoded *trans*-acting regulatory proteins.
- *Antigenic variation*: New antigenic variants of lentiviruses continually arise in infected hosts due to the inaccuracy of reverse transcriptase, which results in a

Table 7.2 Some of the abnormalities of the immune system seen in AIDS

Depletion of CD4$^+$ cells
Lymphopenia
Impaired delayed hypersensitivity reactions
Polyclonal B-cell activation
Decreased humoral response to antigens
Decreased proliferative response to antigens
Decreased proliferative response to T- and B-cell mitogens
Decreased cytotoxicity
Decreased NK cell function
Decreased MHC class II expression
Decreased interferon-γ production
Decreased interleukin-2 production
Decreased monocyte chemotaxis

---

type of antigenic drift. In caprine arthritis encephalitis virus (CAEV) infection of goats, each new antigenic variant has been shown to cause a flare up of inflammatory arthritis. Although HIV does not cause arthritis directly, a similar mechanism may contribute to the decline of immune system in AIDS (see below).

• *The 'Trojan horse' mechanism*: HIV and other lentiviruses escape immune recognition by infecting monocytes, which not only serve as a reservoir for the virus but also help them to spread to other tissues (e.g. the brain) and to other hosts.

It is not clear how much of the pathology of AIDS is due to the virus and how much is actually caused by the immune system. Three of the models which seek to explain how HIV causes immune deficiency are:

Direct cell killing

This was discussed earlier (see Mechanisms of Cellular Injury). It is now generally believed that there is not sufficient virus present in AIDS patients to account for all the damage seen, although killing of CD4$^+$ cells may contribute to the overall pattern of pathogenesis seen in AIDS.

Antigenic diversity

This theory holds that the continual generation of new antigenic variants eventually swamps and overcomes the immune system, leading to its collapse. There is no doubt that new antigenic variants of HIV constantly arise during the long course of AIDS, in a similar way to CAEV infection of goats. It has not been completely established how this might lead to the collapse of the immune system, but it is envisaged that there might be a 'rachet' effect, with each new variant contributing to the slight but irreversible decline in immune function (Figure 7.2). A mathematical model has been constructed which precisely mimics this effect. When primed with all the known data about the state of the immune system during HIV infection, it provides a startlingly accurate depiction of the course of AIDS.

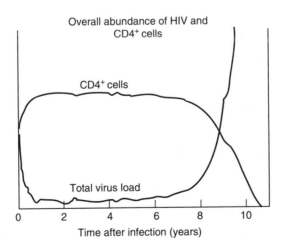

Figure 7.2 Antigenic diversity threshold theory of AIDS pathogenesis. Each line on the upper graph shows the (theoretical) relative abundance of an individual HIV strain (immunologic type) in a single case of AIDS. The lower graph shows the total virus load (i.e. total number of HIV strains) and number of CD4$^+$ cells predicted by a computer model.

The superantigen theory

Superantigens are molecules which short-circuit the immune system, resulting in massive activation of T-cells rather than the usual, carefully controlled response to foreign antigens. It is believed that they do this by binding to both the variable region of the β-chain of the T-cell receptor (Vβ) and to MHC class II molecules, cross-linking them in a non-specific way (Figure 7.3). This results in polyclonal T-cell activation rather than the usual situation where only the few clones of T-cells responsive to a particular antigen presented by the MHC class II molecule are activated. The over-response of the immune system produced results in autoimmunity, as rare clones of T-cells which recognize self antigens are activated, and immunosuppression, as the activated cells subsequently die or are killed by other activated T-cells. It is possible that such superantigens might also induce **apoptosis**, or 'programmed cell killing.' There are two lines of evidence that support this attractive hypothesis. The first is the observation that mice inoculated with T-cells from an allogeneic strain of mice (i.e. mice of a different MHC type) made detectable antibodies against HIV p24 *gag* and gp120 *env* proteins, although neither strain of mice had ever been exposed to HIV. These mice also made anti-MHC antibodies, which would contribute towards immunosuppression. This suggests that

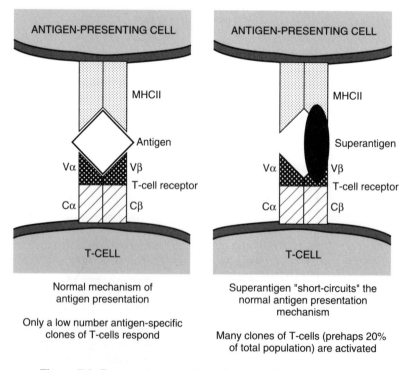

Figure 7.3 Proposed mechanism of action of superantigens.

polyclonal activation of T-cells might be the basis for much of the pathogenesis of AIDS. The second observation is that in some AIDS patients, it has been reported that certain clones of T-cells bearing particular Vβ T-cell receptor rearrangements are depleted or absent. This is precisely what would be expected if some clones of cells were being eliminated by the presence of a superantigen. It has recently been reported that rabies virus N protein is a superantigen which stimulates Vβ8 human T-lymphocytes, so there is a precedent for this mechanism.

Numerous other models have also been suggested and the reader should refer to the references at the end of this chapter for a detailed summary. At the present time the precise basis for the pathogenesis of AIDS is not known. What is clear is that the presence of HIV is necessary for the development of AIDS and that it is vital that the worldwide spread of HIV infection is halted and reversed. Work on developing anti-HIV vaccines is proceeding but because of the complex biology of the virus, is proving to be formidably difficult. A better understanding of the pathogenesis of AIDS and in particular, the role of the immune system in the early stages of the disease, is vital to permit the development of more appropriate therapies for AIDS. In future, this might well involve a two-pronged approach incorporating both an attack on the virus itself plus therapeutic immunomodulation to prevent the decline of the immune system.

Virus-Related Diseases

There are a number of human disease syndromes for which virus infections are believed to be a necessary prerequisite. In some instances, the link between a particular virus and a pathological condition is well established, but it is clear that the pathogenesis of the disease is complex and also involves the immune system of the host. In other cases, the pathogenic involvement of a particular virus is less certain, and in a few instances, rather speculative.

Measles is possibly the most common infectious disease worldwide. The virus causes more than a million deaths per year in children in developing countries. Since the mortality rate of measles is approximately 1 per 1000 cases, this suggests that there must be about 10^9 infections per annum. The normal outcome of measles virus infection is the familiar acute febrile disease, during which the virus spreads throughout the body, infecting many tissues. The vast majority of cases spontaneously recover from the disease without any lasting harm. In a few rare cases (about 1 in 1000), measles may progress to a severe encephalitis. This is still an acute condition which either regresses or kills the patient within a few weeks. There is, however, another, much rarer late consequence of measles virus infection which occurs many months or years after initial infection of the host. This is the condition known as subacute sclerosing panencephalitis (SSPE). Evidence of prior measles virus infection (antibodies or direct detection of the virus) is found in all patients with SSPE, whether they can recall having a symptomatic case of measles or not. In about 1 in 300 000 cases of measles, the virus is not cleared from the body by the immune system, but establishes a persistent infection in the CNS. In this condition, virus replication continues at a low level, but defects in the envelope genes prevent the production of extracellular infectious virus particles. The lack of envelope protein production causes the failure of the immune system to recognize and eliminate infected cells. However, the virus is able to spread directly from cell to cell, bypassing the usual route of infection. It is not known to what extent damage to the cells of the brain is caused directly by virus replication, or whether there is any contribution by the immune system to the pathogenesis of SSPE.

Another well-established case where the immune system is implicated in pathogenesis concerns dengue virus infections. Dengue virus is a flavivirus which is transmitted from one human host to another via mosquitos. The primary infection may be asymptomatic, or may result in dengue fever. Dengue fever is normally a self-limited illness from which patients recover after 7–10 days without further complications. Following primary infections, patients carry antibodies to the virus. Unfortunately, there are four serotypes of dengue virus and the presence of antibody directed against one type does not give cross-protection against the other three. Even worse than this, antibodies can enhance the infection of peripheral blood mononuclear cells by Fc-receptor mediated uptake of antibody-coated dengue virus particles (see Chapter 4). In a few cases, the consequences of dengue virus infection are much more severe than the usual fever. Dengue hemorrhagic fever (DHF) is a life-threatening disease. In the most extreme cases, so much internal hemorrhaging occurs that hypovolemic shock (dengue shock syndrome – DSS) occurs. DSS is frequently fatal. DHF and DSS following primary dengue virus infections occur in approximately 1 in 14 000 and 1 in 500 patients respectively. However, after secondary dengue virus infections, the incidence of DHF is 1 in 90 and DSS 1 in 50, since cross-reactive but non-neutralizing antibodies to the virus are now present. These figures show the problems of cross-infection with different serotypes of dengue virus, and the difficulties which must be faced in developing a safe vaccine against the virus (see Chapter 8).

Another instance where virus vaccines have resulted in increased pathology rather than the prevention of disease is the occurrence of post-vaccination Reye's syndrome. Reye's syndrome is a neurological condition involving acute cerebral edema and occurs almost exclusively in children. It is well known as a rare post-infection complication of a number of different viruses, but most commonly influenza virus. The basis for the pathogenesis of this condition is completely unknown, but some of the most unfortunate cases have followed the administration of experimental influenza virus vaccines. Guillain–Barré syndrome is a similarly mysterious condition in which demyelination of nerves results in partial paralysis and muscle weakness. The onset of Guillain–Barré syndrome usually follows shortly after an acute 'virus-like' infection, but no single agent has ever been firmly associated with this condition. Kawasaki disease is similar to Reye's syndrome in that it occurs in children, but different in that it results in serious damage to the heart. Like Guillain–Barré syndrome, Kawasaki disease appears to follow acute infections. A large number of bacterial and viral pathogens have been suggested to be associated with the induction of Kawasaki disease, but once again, the underlying cause of the pathology is unknown. It would appear that acute infection rather than a particular pathogen is responsible for the onset of these diseases.

In recent years, there has been a search for an agent responsible for a newly diagnosed disease called chronic fatigue syndrome (CFS), or 'post-viral fatigue syndrome.' Unlike the other conditions described above, CFS is a rather ill-defined disease, and is not even recognized by all physicians. Although it is attractive to believe that a virus might be responsible for such conditions as chronic tiredness, so far, this has not been proved. All these various conditions and syndromes illustrate the complexity of viral pathogenesis and show that the direct effects of virus replication and self-inflicted damage resulting from poor control of the immune system are sometimes hard to differentiate.

Cellular Transformation by Viruses

Transformation is a change in the morphological, biochemical, or growth parameters of a cell. Transformation may or may not result in cells able to produce tumors in experimental animals, which is properly known as neoplastic transformation. Therefore, transformed cells do not automatically result in the development of 'cancer.' Carcinogenesis (or more properly, oncogenesis) is a complex, multi-step process, in which cellular transformation may be only the first, although essential, step along the way. Transformed cells have an altered phenotype, which is displayed as one (or more) of the following characteristics:

- *Loss of anchorage dependence*: Normal (i.e. non-transformed) adherent cells such as fibroblasts or epithelial cells require a surface to which they can adhere. In the body, this requirement is supplied by adjacent cells or structures, but *in vitro*, it is met by the glass or plastic vessels in which the cells are cultivated. Some transformed cells lose the ability to adhere to solid surfaces and float free (or in clumps) in the culture medium without loss of viability.
- *Loss of contact inhibition*: Normal adherent cells in culture divide and grow until they have coated all the available surface for attachment. At this point, when adjacent cells are touching each other, cell division stops – the cells do not continue to grow and pile up on top of one another. Many transformed cells have lost this characteristic. Single transformed cells in a culture dish become visible as small thickened areas of growth called 'transformed foci' – clones of cells all derived from a single original cell.
- *Colony formation in semi-solid media*: Most normal cells (both adherent and non-adherent cells such as lymphocytes) will not grow in media that are partially solid due to the addition of substances such as agarose or hydroxymethyl cellulose. However, many transformed cells will grow under these conditions, forming colonies since movement of the cells is restricted by the medium.
- *Decreased requirements for growth factors*: All cells require multiple factors for growth. In a broad sense, these include compounds such as ions, vitamins, and hormones which cannot be manufactured by the cell. More specifically, it includes regulatory peptides such as epidermal growth factor (EGF) and platelet-derived growth factor (PDGF) which regulate the growth of cells. These are potent molecules which have powerful effects on cell growth. Some transformed cells may have decreased or may even have lost their requirement for particular factors. The production by a cell of a growth factor which is required from its own growth is known as **autocrine** stimulation and is one route by which cells may be transformed.

Cell transformation is a single hit process, i.e. a single virus transforms a single cell. All or part of the virus genome persists in the transformed cell, and is usually (but not always) integrated into the host cell chromatin. Transformation is usually accompanied by continued expression of a limited repertoire of virus genes, or rarely by productive infection. Virus genomes found in transformed cells are frequently replication-defective and contain substantial deletions.

Transformation is mediated by proteins encoded by **oncogenes**. These regulatory genes can be grouped in several ways, for example, by their origins, biochemical functions, or subcellular locations. Oncogenes are involved in the control of cell growth and are found in all cells. About 25 such genes have now been identified, and these normal cellular sequences are known as cellular oncogenes or *c-oncs*. Cell-transforming viruses may have RNA or DNA genomes, but all have at least a DNA stage in their replication cycle, i.e. the only RNA viruses capable of cell transformation are the retroviruses (Table 7.3). Certain retroviruses carry homologs of *c-oncs* derived originally from the cellular genes and known as *v-oncs*. In contrast, the oncogenes of cell-transforming DNA viruses are unique to the virus genome – there are no homologous sequences present in normal cells. Oncogenes can be grouped by their common biochemical functions:

Table 7.3 Known oncogenes transduced by retroviruses

Gene	Protein function	Type of tumour
abl	Tyrosine kinase	Pre-B-cell leukemia/sarcoma
akt	?	T-cell lymphoma/sarcoma
crk	Activator of tyrosine kinase	
erbA	Thyroid hormone receptor	
erbB	Epidermal growth factor receptor	Erythroleukemia/fibrosarcoma
ets	Nuclear protein	
fes/fps	Tyrosine kinase	Sarcoma
fgr	Tyrosine kinase	Sarcoma
fms	Tyrosine kinase; macrophage colony stimulating factor receptor	Sarcoma
fos	Nuclear transcription factor	Osteosarcoma
jun	AP-1 transcription factor	Fibrosarcoma
kit	Tyrosine kinase	Sarcoma
mil/raf	Serine/threonine kinase	Sarcoma
mos	Serine/threonine kinase	Sarcoma
myb	Nuclear protein	Myeloblastosis
myc	Nuclear protein	Sarcoma/myelocytoma/carcinoma
H-*ras*	G protein	Sarcoma/erythroleukemia
K-*ras*	G protein	Sarcoma/erythroleukemia
rel	Nuclear protein	Reticuloendotheliosis
ros	Tyrosine kinase	Sarcoma
sea	Tyrosine kinase	Sarcoma/leukemia
sis	Platelet-derived growth factor	Sarcoma
ski	Nuclear protein	Carcinoma
src	Tyrosine kinase	Sarcoma
yes	Tyrosine kinase	Sarcoma

- Extracellular growth factors – homologs of normal growth factors.
- Membrane receptors – responsible for the capture of extracellular signals.
- Tyrosine kinases – associated with the inner surface of the cell membrane.
- G proteins involved in signal transduction.
- Cytoplasmic protein kinases.
- Transcription factors (see Chapter 5).
- Nuclear hormone receptors.

The function of each class of oncogene product determines their cellular location (Figure 7.4). Several classes of oncogenes are associated with the process of signal transduction, i.e. the transfer of information derived from the binding of extracellular ligands to cellular receptors to the nucleus (Figure 7.5). Many of the kinases in these groups have a common type of structure with conserved functional domains representing the hydrophobic transmembrane and hydrophilic intracellular kinase regions (Figure 7.6). These proteins are associated with the cell membranes or are present in the cytoplasm. Other classes of oncogenes located in the nucleus are normally involved with the control of the cell cycle (Figure 7.7). The products of these genes overcome the restriction between the G_1 and S phases of the cell cycle, which is the key control point in preventing uncontrolled cell division.

Some viral oncogenes are not sufficient on their own to produce a fully transformed phenotype in cells. However, in some instances, they may cooperate with another oncogene of complementary function to produce a fully transformed phenotype, for example, the adenovirus E1A gene plus the cellular *ras* gene transforms NIH3T3 cells (a mouse fibroblast cell line). This further underlines the fact that transformation (and oncogenesis) is a complex, multi-step process.

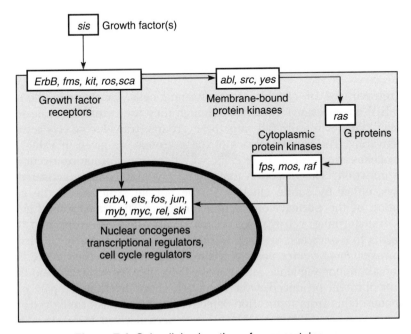

Figure 7.4 Subcellular location of oncoproteins.

Figure 7.5 Cellular mechanism of signal transduction.

Cell transformation by retroviruses

Not all retroviruses are capable of transforming cells, for example, the lentiviruses such as HIV do not transform cells, although they are cytopathic. The retroviruses which can transform cells fall into three groups: transducing, *cis*-activating, and *trans*-activating. The characteristics of these groups are given in Table 7.4.

If oncogenes are present in all cells, why does transformation occur as a result of virus infection? The reason is that oncogenes may become activated in one of two ways, either by subtle changes to the normal structure of the gene, or by interruption of the normal control of expression. The transforming genes of the acutely transforming retroviruses (*v-oncs*) are derived from and are highly homologous to *c-oncs* and are believed to have been transduced by viruses (Table 7.3). However, most *v-oncs* possess slight alterations from their *c-onc* progenitors. Many contain minor sequence alterations which alter the structure and the function of the oncoprotein produced. Others contain short deletions of part of the gene. Most oncoproteins from replication-defective, acutely transforming retroviruses are fusions proteins, containing additional sequences derived from virus genes, most commonly viral *gag* sequences at the amino terminus of the protein. These additional

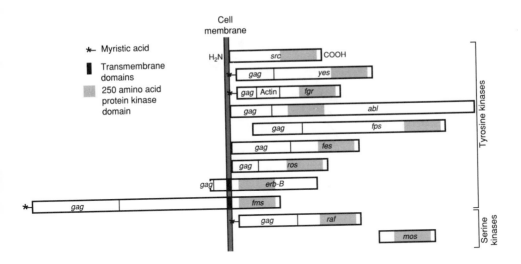

Figure 7.6 The family of retrovirus protein kinases involved in cell transformation. Many of these molecules are fusion proteins containing amino-terminal sequences derived from the *gag* gene of the virus. Most of this type contain the fatty acid myristate which is added to the amino terminus of the protein after translation and which links the protein to the inner surface of the host cell cytoplasmic membrane. In a number of cases, it has been shown that this post-translational modification is essential to the transforming action of the protein.

sequences may alter the function or the cellular localization of the protein and these abnormal attributes result in transformation.

Alternatively, viruses may result in abnormal expression of an unaltered oncoprotein. This might be either the overexpression of an oncogene under the control of a virus promoter rather than its normal promoter in the cell, or it may be the inappropriate temporal expression of an oncoprotein which disrupts the cell cycle. Chronic transforming retrovirus genomes do not contain oncogenes. These viruses activate *c-oncs* by a mechanism known as insertional activation. A provirus which integrates into the host cell genome close to a *c-onc* sequence may indirectly activate the expression of the gene in a way analogous to that in which *v-oncs* have been activated by transduction (Figure 7.8). This can occur if the provirus is integrated upstream of the *c-onc* gene, which might be expressed via a read-through transcript of the virus genome plus downstream sequences. However, insertional activation can also occur when a provirus integrates downstream of a *c-onc* sequence or upstream but in an inverted orientation. In these cases, activation results from enhancer elements in the virus promoter (see Chapter 5). These can act even if the provirus integrates at a distance of several kilobases from the *c-onc* gene. The best-known examples of this phenomenon occur in chickens, where

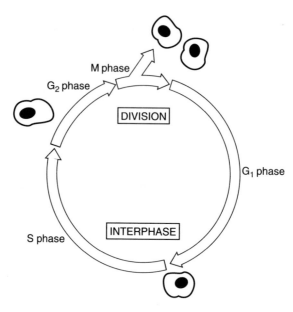

Figure 7.7 Schematic diagram showing the phases of the eukaryotic cell cycle.

Table 7.4 Cell-transforming retroviruses

Virus type	Time to tumor formation	Efficiency of tumor formation	Type of oncogene
Transducing (acutely transforming)	Short (e.g. weeks)	High – up to 100%	*c-onc* transduced by virus, i.e. *v-onc* present in virus genome
Cis-activating (chronic transforming)	Intermediate (e.g. months)	Intermediate	*c-onc* in cell genome activated by provirus insertion – no oncogene present in virus genome
Trans-activating	Long (e.g. years)	Low (< 1%)	Activation of cellular genes by *trans*-acting virus proteins

insertion of avian leukosis virus (ALV) activates the *myc* gene, and in mice, where mouse mammary tumor virus (MMTV) insertion activates the *int* gene.

Transformation by the third class of retroviruses operates by quite a different mechanism. Human T-cell leukemia virus (HTLV) and related animal viruses encode a transcriptional activator protein in the virus *tax* gene (see Chapter 5). The tax protein acts in *trans* to stimulate transcription from the virus LTR. It is believed that the protein also activates transcription of some cellular genes, notably those encoding the T-cell growth factor interleukin-2 (IL-2) and its receptor (IL-2R). This could result in the formation of a cycle of positive feedback involving autocrine

Figure 7.8 Mechanisms by which cellular oncogenes can be transcriptionally activated by retrovirus insertional mutagenesis.

stimulation of T-cells by these two genes. However, HTLV oncogenesis, i.e. the formation of a leukemic tumor, has a latent period of some 20–30 years. Therefore, cell transformation (which can be mimicked *in vitro*) and tumor formation (which cannot) are not one and the same – additional events required for the development of leukemia. It is thought that chromosomal abnormalities which may occur in the population of HTLV-transformed cells are also required to produce a malignant tumor, although because of the difficulties of studying this lengthy process, this is not completely understood.

Cell transformation by DNA viruses

In contrast to the oncogenes of retroviruses, the transforming genes of DNA tumor viruses have no cellular counterparts. Several families of DNA viruses are capable of transforming cells (Table 7.5). In general terms, the functions of their oncoproteins are much less diverse than those encoded by retroviruses. They are mostly nuclear proteins involved in the control of DNA replication which directly affect the cell cycle. They achieve their effects by interacting with cellular proteins which normally appear to have a negative regulatory role in cell proliferation. Two of the most important cellular proteins involved are known as p53 and p105-*RB*.

p53 was originally discovered by virtue of the fact that it forms complexes with SV40 T-antigen. It is now known that it also interacts with other DNA virus oncoproteins, including those of adenoviruses and papillomaviruses. The gene encoding p53 is mutated or altered in the majority of tumors implying that loss of

Table 7.5 Transforming proteins of DNA tumor viruses. These proteins are uniquely virus encoded and, unlike (most) retrovirus oncogenes, have no cellular counterparts

Virus	Protein	Cellular localization
Polyomaviruses		
Mouse polyomavirus	Large T-antigen	Nucleus
	Middle T-antigen	Cell membrane
	Small T-antigen	Nucleus/cytoplasm
SV40	Large T-antigen	Nucleus
	Small T-antigen	Nucleus/cytoplasm
Adenoviruses	E1A	Nucleus
	E1B	Nucleus
Papillomaviruses	E5	Cell membrane
	E6	Nucleus
	E7	Nucleus

the normal gene product is associated with the emergence of malignantly transformed cells. Tumor cells, when injected with the native protein *in vitro* show a decreased rate of cell division and decreased tumorigenicity *in vivo*. Transgenic mice which do not possess an intact p53 gene are developmentally normal but are susceptible to the formation of spontaneous tumors. Therefore, it is clear that p53 plays a central role in controlling the cell cycle. It is believed to be a tumor suppressor or 'anti-oncogene' and has been called 'the guardian of the genome.' p105-*RB* was discovered when it was noticed that the gene which encodes this protein is always damaged or deleted in a tumor of the optic nerve known as retinoblastoma. Therefore, the normal function of this gene is also thought to be that of a tumor suppressor. It is not yet understood how these proteins act to prevent the formation of tumors. One idea is that they may monitor the cell genome for genetic damage. If damage to the DNA has occurred, they may block the cell cycle at the G_1 arrest stage, allowing time for DNA repair to be carried out. In the absence of these proteins, cell division proceeds and the damaged DNA is copied into other cells, resulting eventually in the formation of a malignant tumor. Since DNA tumor viruses carry out DNA replication in the nucleus of their host cells, they must neutralize the cellular tumor suppressor proteins by forming complexes with virus proteins (see below).

SV40 T-antigen is one of the known virus proteins which binds p53. Chapter 5 describes the role of large T-antigen in the regulation of SV40 transcription. Infection of cells by SV40 or other polyomaviruses can result in two possible outcomes:

• Productive (lytic) infection
• Non-productive (abortive) infection.

The outcome of infection appears to be determined primarily by the cell type infected; for example, mouse polyomavirus establishes a lytic infection of mouse cells but an abortive infection of rat or hamster cells, while SV40 shows lytic infection of monkey cells but abortive infection of mouse cells. However, in addition to transcription, T-antigen is also involved in genome replication. SV40 DNA replication is initiated by binding of large T-antigen to the origin region of the genome (Figure 5.12). The function of T-antigen is controlled by phosphorylation, which decreases the ability of the protein to bind to the SV40 origin.

The SV40 genome is very small and does not encode all of the information necessary for DNA replication. Therefore, it is essential for the host cell to enter S phase, when cell DNA and the virus genome are replicated together. Protein–protein interactions between T-antigen and DNA polymerase α directly stimulate replication of the virus genome. The precise regions of the T-antigen involved in binding to DNA, DNA polymerase α, p53, and p105-*RB* are all known (Figure 7.9). Inactivation of tumor suppressor proteins bound to T-antigen causes G_1-arrested cells to enter S phase and divide and this is the mechanism which results in transformation. However, the frequency with which abortively infected cells are transformed is low (about 1×10^{-5}). Therefore, the function of T-antigen is to alter the cellular environment to permit virus DNA replication. Transformation is a rare and accidental consequence of the sequestration of tumor suppressor proteins.

The immediate early proteins of adenoviruses are analogous in many ways to SV40 T-antigen. E1A is a *trans*-acting transcriptional regulator of the adenovirus early genes (see Chapter 5). Like T-antigen, the E1A protein binds to p105-*RB*, inactivating the regulatory effect of this protein, permitting viral DNA replication, and accidentally stimulating cellular DNA replication (see above). E1B binds p53 and reinforces the effects of E1A. The combined effect of the two proteins can be seen in the phenotype of cells transfected with DNA containing these genes:

DNA:	*Cell phenotype*:
E1A	Immortalized but morphologically unaltered; not tumorigenic in animals.
E1B	Not transformed.
E1A + E1B	Immortalized *and* morphologically altered; tumorigenic in animals.

Papillomaviruses also transform cells, but the mechanism of transformation is less well understood than in polyomaviruses or adenoviruses. Human papillomavirus (HPV) genital infections are very common, occurring in more than 50% of young, sexually active adults, and are usually asymptomatic. Certain serotypes of HPV (e.g. HPV-6, HPV-11) are associated with a low risk of subsequent development of

Figure 7.9 Regions of SV40 T-antigen involved in protein–protein interactions. Other functional domains of the protein involved in virus DNA replication are also shown, e.g. the helicase, ATPase, and nuclear location signal (NLS) domains.

anogenital cancers such as cervical carcinoma, after an incubation period of 15–25 years. According to some reports, 85–90% of such tumors contain transcriptionally active HPV-16 or HPV-18. Of the 63 HPV types currently recognized, 25 are associated with a high risk of tumor formation (e.g. HPV-16, 18, 31, 33, etc.). Once again, transformation is mediated by the early gene products of the virus. However, the transforming proteins appear to vary from one type of papillomavirus to another as shown in Table 7.6.

There is currently some confusion about the normal functions and mechanism of cell transformation by these proteins. The E6 protein is known to bind p53 and to target it for accelerated degradation, decreasing the intracellular level of the protein. The E7 protein has structural and functional similarities (e.g. binding to p105-*RB*) with adenovirus E1A. In general terms, it appears that two or more early proteins often cooperate to give a transformed phenotype. Although some papillomaviruses can transform cells on their own (e.g. BPV-1), others appear to require the cooperation of an activated cellular oncogene (e.g. HPV-16/*ras*). More confusingly, in most cases, all or part of the papillomavirus genome, including the putative transforming genes, is maintained in the tumor cells, whereas in some cases (e.g. BPV-4), the virus DNA may be lost after transformation, which may indicate a possible hit-and-run mechanism of transformation. Different papillomaviruses appear to use slightly different mechanisms to achieve genome replication and consequently, cell transformation may proceed via a slightly different route. It is imperative that a better understanding of these processes is obtained. There is no positive evidence that adenoviruses or polyomaviruses are involved in the formation of human tumors. In contrast, the evidence that papillomaviruses may be involved in the formation of malignant penile and cervical carcinomas is now strong.

Viruses and Cancer

There are numerous examples of viruses which cause tumors in experimental animals. This stimulated a long search for viruses which might be the cause of cancer in man. For many years, this search was unsuccessful, so much so that a few scientists categorically stated that viruses did not cause human tumors. Like all rash statements, this was wrong. We currently know of at least three viruses which are associated with the formation of tumors in infected people. However,

Table 7.6 Transforming proteins of papillomaviruses

	Cell transformation with gene products		
Virus	E5	E6	E7
Bovine papillomavirus (BPV-1)	+	+	−
Human papillomavirus (HPV-16)	+	+ (binds p53)	+ (binds p105-*RB*)
Human papillomavirus (HPV-8)	−	+	−

the relationship between virus infection and tumorigenesis is indirect and complex. The role of the HTLV tax protein in leukemia has already been described (see Cell Transformation by Retroviruses, page 180). The evidence that papillomaviruses may be involved in human tumors is also growing. There are almost certainly many more viruses which cause human tumors, but the remainder of this chapter will describe two examples which have been intensively studied.

Epstein–Barr virus (EBV) was first identified in 1964 in a lymphoblastoid cell line derived from an African patient with Burkitt's lymphoma. In 1962, Dennis Burkitt described a highly malignant lymphoma whose distribution in Africa paralleled that of malaria. Burkitt recognized that this tumor was rare in India, but occurred in Indian children living in Africa, and therefore looked for an environmental cause. Initially, he thought that the tumor might be caused by a virus spread by mosquitoes (which is wrong). The association between EBV and Burkitt's lymphoma is not entirely clear cut:

- EBV is widely distributed worldwide but Burkitt's lymphoma is rare.
- EBV is found in many cell types in Burkitt's lymphoma patients, not just in the tumor cells.
- Rare cases of EBV-negative Burkitt's lymphoma are sometimes seen in countries where malaria is not present, suggesting there may be more than one route to this tumor.

EBV has a dual cell tropism for human B-lymphocytes (generally a non-productive infection) and epithelial cells, in which a productive infection occurs. The usual outcome of EBV infection is polyclonal B-cell activation and a benign proliferation of these cells, which is frequently asymptomatic, but sometimes produces a relatively mild disease known as infectious mononucleosis or glandular fever. In 1968, it was shown that EBV could efficiently transform (i.e. immortalize) human B-lymphocytes *in vitro*. The mechanism by which EBV transforms cells is still not precisely understood, but the observation that it can do so clearly strengthens the case that it is involved in the formation of tumors. There is now epidemiological and/or molecular evidence that EBV infection is associated with at least four human tumors:

- Burkitt's lymphoma.
- Nasopharyngeal carcinoma (NPC), a highly malignant tumor seen most frequently in China. There is a strong association between EBV and NPC – unlike Burkitt's lymphoma, the virus has been found in all the tumors which have been studied. Environmental factors, such as the consumption of nitrosamines in salted fish, are also believed to be involved in the formation of NPC (cf. the role of malaria in the formation of Burkitt's lymphoma).
- B-cell lymphomas in immunosuppressed individuals, e.g. AIDS patients.
- Some clonal forms of Hodgkin's disease.

There are three possible explanations to explain the link between EBV and Burkitt's lymphoma:

(1) EBV immortalizes a large pool of B-lymphocytes. Concurrently, malaria causes T-cell immunosuppression. There is thus a large pool of target cells in which

a third event (e.g. a chromosomal translocation) results in the formation of a malignantly transformed cell. Most Burkitt's lymphoma tumors contain translocations involving chromosome 8, resulting in activation of c-*myc* gene, which supports this hypothesis.

(2) Malaria results in polyclonal B-cell activation. EBV subsequently immortalizes a cell containing a pre-existing c-*myc* translocation. This mechanism would be largely indistinguishable from the above.

(3) EBV is just a passenger virus! Burkitt's lymphoma occurs in Europe and North America, although it is very rare in these regions, but 85% of these patients are not infected with EBV, which implies that there are other causes for Burkitt's lymphoma.

Although it has not been formally proved, it nevertheless seems likely that either (1) and/or (2) are the true explanations for the origin of Burkitt's lymphoma.

Another case where a virus appears to be associated with the formation of a human tumor is that of hepatitis B virus (HBV) and hepatocellular carcinoma (HCC). Hepatitis is an inflammation of the liver and as such is not a single disease. Because of the central role of the liver in metabolism, many virus infections may involve the liver. However, there are at least seven viruses which specifically seem to infect and damage hepatocytes. No two of these belong to the same family (see Chapter 8). HBV is the prototype member of the family *Hepadnaviridae* and causes the disease formerly known as 'serum hepatitis.' This disease was distinguished clinically from 'infectious hepatitis' (caused by other types of hepatitis virus) in the 1930s. HBV infection used to be the result of inoculation with human serum (e.g. blood transfusions, organ transplants), but is still common among intravenous drug abusers where it is spread by the sharing of needles and syringes. However, the virus is also transmitted sexually, by oral ingestion, and from mother to child, accounting for familial clusters of HBV infection. All blood, organ, and tissue donations in developed countries are now tested for HBV and risk of transmission is extremely low. The virus does not replicate in tissue culture and this has seriously hindered investigations into its pathogenesis. HBV infection has three possible outcomes:

- An acute infection followed by complete recovery and immunity from reinfection (> 90% of cases).
- Fulminant hepatitis, developing quickly and lasting a short time, causing liver failure and a mortality rate of approximately 90% (< 1% of cases).
- Chronic infection, leading to the establishment of a carrier state with virus persistence (about 10% of cases).

There are approximately 250 million chronic HBV carriers worldwide. The total population of the world is approximately 5.5 billion, therefore about 5% of the world's population is infected with HBV. All of these chronic carriers of the virus are at 100–200 times the risk of non-carriers of developing HCC. HCC is a rare tumor in the west, where it represents < 2% of fatal cancers. Most cases which do occur in the west are alcohol-related and this is an important clue to the pathogenesis of the tumor. However, in South-East Asia and in China, HCC is the most common fatal cancer, causing about half a million deaths every year. The virus might cause the formation of the tumor by three different pathways: direct activation of a cellular oncogene(s), *trans*-activation of a cellular oncogene(s), or indirectly via tissue regeneration (Figure 7.10). As with EBV and Burkitt's lymphoma, the relationship between HBV and HCC is not clear cut:

- Cirrhosis (a hardening of the liver which may be the result of infections or various toxins, e.g. alcohol) appears to be a prerequisite for the development of HCC. It would appear that chronic liver damage induces tissue regeneration and that faulty DNA repair mechanisms result eventually in malignant cell transformation.
- A number of co-factors, such as aflatoxins and nitrosamines, can induce HCC-like tumors in experimental animals without virus infection. Therefore, such substances may also be involved in human HCC (cf. nitrosamines and NPC, above).
- There is no consistent evidence for the integration of the HBV genome or even the persistence of particular HBV genes (e.g. the X gene, which encodes a *trans*-activator protein functionally analogous to the HTLV tax protein) in tumor cells.

It is possible that all the mechanisms shown in Figure 7.10 might operate *in vivo*. The key risk factor is the development of a chronic as opposed to an acute HBV infection. This in itself is determined by a number of other factors:

- *Age*: The frequency of chronic infections declines with increasing age at the time of infection.
- *Sex*:
  Chronic infection: male/female ratio: 1.5 : 1.
  Cirrhosis: male/female ratio 3 : 1.
  HCC: male/female ratio 6 : 1.

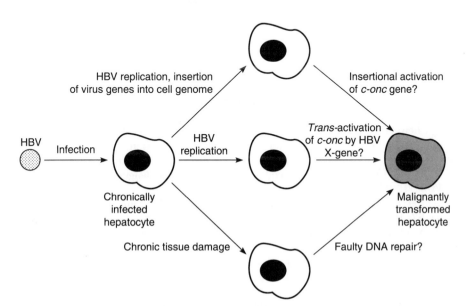

Figure 7.10 Possible mechanisms of hepatocellular carcinoma formation due to hepatitis B virus infection.

- *Route of infection*: Oral or sexual infections give rise to less cases of chronic infection than serum infection.

Until there is a much better understanding of the pathogenesis and normal course of HBV infection it is unlikely that the reasons for these differences will be understood. There may be a happy ending to this story. Safe and effective vaccines which prevent HBV infection are now available. If these are widely used in the areas of the world where HBV infection is endemic, this will in the future prevent half a million deaths from HCC each year.

Summary

Viral pathogenesis is a complex, variable, and relatively rare state. Like the course of a virus infection, pathogenesis is determined by the balance between host and viral factors. Not all of the pathogenic symptoms seen in virus infections are caused directly by the virus – the immune system also plays a part in causing cell and tissue damage. Viruses can transform cells so that they continue to grow indefinitely. In some but not all cases, this can lead to the formation of tumors. There are at least a few well-established cases where certain viruses provoke human tumors and possibly many others which we do not yet understand. The relationship between the virus and the formation of the tumor is not a simple one, but the prevention of infection undoubtedly reduces the risk of tumor formation.

Further Reading

Mims, C.A. (1987). *The Pathogenesis of Infectious Disease* (3rd edition). Academic Press, London.
Roitt, I. (1988). *Essential Immunology* (6th edition). Blackwell Scientific, Oxford.

Bibliography

Fan, H.Y., Chen, I.S.Y., Rosenberg, N. and Sugden, W. (Eds) (1991). *Viruses That Affect the Immune System*. American Society for Microbiology, Washington, D.C.
Johnson, H.M. *et al.* (1992). Superantigens in human disease. *Scientific American* **266**: 20–6.
Kurane, I. *et al.* (1991). Antibody-dependent enhancement of dengue virus infection. *Rev. Med. Virol.* **1**: 211–21.
Levy, J.A. (1993). Pathogenesis of Human Immunodeficiency Virus infection. *Microbiological Reviews* **57**: 183–289.
Nowak, M.A. *et al.* (1991). Antigenic diversity thresholds and the development of AIDS. *Science* **254**: 963–9.
Spector, S. *et al.* (Eds) (1989). *Virus-Induced Immunosupression*. Plenum Press, New York.
Temin, H. (1988). Mechanisms of cell killing/cytopathic effects by non-human retroviruses. *Rev. Infect. Dis.* **10**: 399–405.
Tiollais, P. and Buendia, M. (1991). Hepatitis B virus. *Scientific American*, April, 48–54.

Chapter 8

Novel Infectious Agents

What is the minimum genome size necessary to sustain an infectious agent? Could a virus with a genome of 1700 nucleotides exist? Could an agent with a genome of 240 nucleotides exist? Could an apparently infectious agent without any genome at all exist? Perhaps the first two alternatives might be possible, but the idea of an infectious agent without a genome seems ridiculous. Nevertheless, strange as it may seem, such agents do indeed exist and cause disease in animals (including man) and plants. By reading the earlier chapters of this book, readers should have a grasp of the idea of what constitutes a conventional virus in terms of structure, replication, and disease-causing potential. But what makes a virus constitute a 'novel' agent? Does this term just apply to viruses that have never been encountered before, or should it include previously known viruses which appear to have changed their behavior?

This chapter will describe and attempt to explain current understanding of a number of agents which meet the above criteria. Unlike the previous chapters, much of the information presented here is not the result of many years of study, but is derived from very recent investigations into these unconventional infectious agents.

Emergent Viruses

In the last few decades there have been massive and unexpected epidemics caused by certain viruses. For the most part, these are not caused by completely new (i.e. previously unknown) viruses, but by viruses which were well known in the geographical areas in which they have suddenly caused epidemic outbreaks of disease. Such viruses are known as **emergent viruses**. There are numerous examples of such viruses which appear to have mysteriously altered their behavior with time.

One of the better known examples of this phenomenon is poliovirus. It is known that poliovirus and poliomyelitis have existed in human populations for at least 4000 years. However, for most of this time, the pattern of disease was **endemic**, i.e. a low, continuous level of infection in a particular geographical area, rather than **epidemic**. During the first half of this century, the pattern of occurrence of poliomyelitis in Europe, North America, and Australia changed to an epidemic one, with vast annual outbreaks of infantile paralysis. Although we do not have samples of poliovirus from earlier centuries, the clinical symptoms of the disease give no reason to believe that the virus changed substantially. Why then did the pattern of disease change so dramatically? It is believed that the reason is as follows. In rural communities with primitive sanitation facilities, poliovirus circulated freely. Serological surveys in similar contemporary situations reveal that more than 90% of children of 3 years of age have antibodies to at least one of the three serotypes of poliovirus. Even the most virulent strains of poliovirus cause 100–200 subclinical infections for each case of paralytic poliomyelitis seen. In such communities, infants experience subclinical immunizing infections whilst still protected by maternal antibodies – a form of natural vaccination. Therefore, in these circumstances, the relatively few cases of paralysis and death which do occur are likely to be overlooked, especially in view of high infant mortality rates. During the nineteenth century, industrialization and urbanization changed the pattern of poliovirus transmission. Dense urban populations and increased traveling afforded opporutnities for rapid transmission of the virus. In addition, improved sanitation broke the natural pattern of virus transmission. Children were likely to encounter the virus for the first time at a later age and without the protection of maternal antibodies. Therefore, these children were at far greater risk when they did eventually become infected and it is believed that these social changes account for the altered pattern of disease. Fortunately, the widespread use of poliovirus vaccines has since brought the situation under control in industrialized countries, although poliomyelitis continues to be a major problem in developing countries.

There are many examples of the epidemic spread of viruses caused by movement of human populations. Measles and smallpox were not known to the ancient Greeks. Both of these viruses are maintained by direct person-to-person transmission and have no known alternative hosts. Therefore, it has been suggested that it was not until human populations in China and the Roman empire reached a certain critical density that these viruses were able to propagate in an epidemic pattern and cause recognizable outbreaks of disease. Prior to this time, the few cases that did occur could easily have been overlooked. Smallpox reached Europe from the Far East in AD 710 and in the eighteenth century, achieved plague proportions – five reigning European monarchs died from smallpox during this century. However, the worst effects occurred when these viruses were transmitted to the New World. Smallpox was (accidentally) transferred to the Americas by Hernando Cortez in 1520. In the next two years, 3.5 million Aztecs died from the disease and the Aztec empire was decimated by disease rather than by conquest. Although not as highly pathogenic as smallpox, epidemics of measles subsequently finished off the Aztec and Inca civilizations. More recently, the first contacts with isolated groups of Inuit and tribes in New Guinea and South America have had similarly devastating results, although on a smaller scale. These historical incidents illustrate the way in which a known virus can suddenly cause illness and death on a catastrophic scale following a change in human behavior.

Measles and smallpox viruses are transmitted exclusively from one human host to another. For viruses with more complex cycles of transmission, for example, those with secondary hosts and insect vectors, the problems are magnified and control of infection becomes much more difficult (Figure 8.1). This is particularly

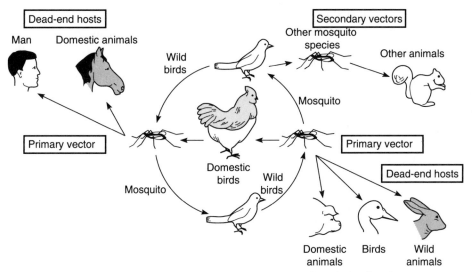

Figure 8.1 Complex transmission pattern of an 'arbovirus.'

true of the families of viruses known collectively as 'arboviruses', i.e. arthropod-borne viruses, (togaviruses, flaviviruses, bunyaviruses, and arenaviruses). As man's territory has expanded, this has increasingly brought people into contact with the type of environment where these viruses are found – warm, humid, vegetated areas where insect vectors occur in high densities, such as swamps and jungles. The classic example of this is the mortality caused by yellow fever virus during the building of the Panama Canal at the end of the nineteenth century.

More recently, the increasing pace of ecological alteration in tropical areas has resulted in massive epidemics of diseases such as dengue fever. This disease was first described in 1780 and by 1906 it was known that the virus was transmitted by mosquitoes. The virus was isolated in 1944. Therefore, this is not a new virus, but the frequency of dengue virus infection has increased dramatically in the last 50 years due to factors such as expanding human populations in tropical areas, increases in *Aedes aegyptii* mosquitoes (the main vector), and increases in worldwide air travel. Some outbreaks of dengue fever have involved more than a million cases with attack rates of up to 90% of the population. In the Americas (North and South) in 1986 the total mortality caused by dengue virus was estimated to be approximately two million cases.

Of more than 520 arboviruses known, at least 100 are pathogenic for man and perhaps 20 would meet the criteria for 'emergent viruses.' Other viruses which are a particular cause for concern at the present time include Rift Valley fever virus and the Hantaan virus group (bunyaviruses). Rift Valley fever virus has caused repeated epidemics in sub-Saharan Africa during the last decade. In one of these, the outbreak has been related to the construction of a dam on the Senegal River which increased mosquito populations. Hantaviruses are transmitted direct from rodents to man and cause millions of cases of hemorrhagic fever each year in many parts of the world, including the USA. Junin virus (an arenavirus) is also transmitted from rodents to man and is a major problem in South America.

Attempts to control these diseases rely on twin approaches involving both the control of insect vectors responsible for transmission of the virus to man and the development of vaccines to protect human populations. However, both of these approaches present considerable difficulties, the former in terms of avoiding environmental damage and the latter in terms of understanding viral pathogenesis and developing appropriate vaccines (see discussion of dengue virus pathogenesis in Chapter 7, page 176).

In some cases, emergent viruses are, however, genuinely novel, i.e. have not been previously identified. The most obvious examples of this are human T-cell leukemia virus (discovered in 1981) and human immunodeficiency virus (discovered in 1983). However, in addition to these, there are a number of other viruses which have come to prominence in recent years. Although not as dramatic as the causative agent of AIDS, there are a number of relatively unknown viruses which cause human gastroenteritis. The taxonomy of these viruses is presently rather confused, and because they do not replicate *in vitro* or in most experimental animals, they have been little studied until recently. The groups of viruses include:

- *Caliciviruses*: There are three well-characterized caliciviruses: vesicular exanthema of swine virus (VESV), San Miguel sea lion virus (SMSLV), and feline calicivirus (FCV). In addition, there are many others identified by electron microscopy which infect insects, reptiles, fish, birds, pigs, cattle, dogs, chimpanzees, and humans. Human caliciviruses (HCV) are commonly identified by electron microscopy in fecal specimens from patients with diarrhea. At least five distinct serotypes of calicivirus have been identified so far. These viruses are widely distributed worldwide and infection is very common, especially in children – most adults appear to have been infected and may or may not be immune to reinfection. Calicivirus infection is associated with diarrhea or vomiting lasting 1–11 days. Evidence of the pathogenic properties of these viruses comes from the fact that contacts of infected individuals often develop the same symptoms and that contacts can be shown in human volunteer studies to excrete an apparently similar virus.
- *Astroviruses*: This group of viruses was first described in 1975. The name of the family is derived from their appearance in electron microscopy – a smooth or slightly indented outer electron-dense shell with an inner, negatively staining 5 or 6 pointed star-shaped core. Clinically, these viruses cause similar symptoms to caliciviruses. Astrovirus infection is very common worldwide. Astroviruses have been isolated from birds, cats, dogs, pigs, sheep, cows, and humans. There are at least five human astrovirus serotypes. In the UK, approximately 75% of the adult population has antibodies against these viruses, indicating prior infection.
- *Small round structured viruses (SRSVs)*: These viruses are even less well characterized than the above groups. The first to be identified (in 1972) was the Norwalk agent, named after a town in Ohio, USA. There are a number of similar viruses all named after their place of isolation which have been partially characterized, e.g. Taunton, Hawaii, Snow Mountain, etc. There are at least 14 different serotypes known. The epidemiology and clinical symptoms associated with SRSV infections resemble those of caliciviruses and astroviruses and it is

possible that they may eventually turn out to be members of these two families. The true prevalence of these viruses is hard to determine, but they may be very common – some reports suggest they are the commonest cause of diarrhea in the UK.

Diarrhea is clearly not as dramatic an illness as AIDS. However, the above groups of viruses are of great economic importance, causing millions of lost working days each year, as well as much discomfort. One well-documented source of infection is the consumption of shellfish (polluted by sewage) – and, therefore, they also have economic consequences for fishermen and the food industry. Such viruses often cause mini-epidemics in families, hospital wards, etc. and are potentially very dangerous to seriously ill hospital patients. More importantly, these viruses contribute to the massive mortality caused by infantile diarrhea in developing countries and are responsible for uncounted millions of deaths each year.

In recent years, two new human herpesviruses have been discovered. Human herpesvirus 6 (HHV-6) was isolated in 1986 in lymphocytes from patients with lymphoreticular disorders. This virus has now been recognized as causing a very common human infection and discovery of HHV-6 solved a longstanding mystery. Primary infection with this virus in childhood causes 'roseola infantum' or 'fourth disease', a common childhood rash whose cause was previously unknown. HHV-6 antibody titres are highest in children and decline with age. The clinical consequences of HHV-6 infection appear to be mild and the virus is not known to be associated with any severe illness. Human herpesvirus 7 (HHV-7) was first isolated from human CD4$^+$ cells in 1990. Little is known about this virus at present, although it does not appear that it causes any severe illness. However, this opinion might have to be revised if new facts come to light in the future.

Although many different virus infections may involve the liver, there are at least seven viruses which seem specifically to infect and damge hepatocytes. No two of these belong to the same family! The identification of these viruses has been a long story:

- *Hepatitis A virus (HAV)*: The virus was first isolated by Purcell in 1973. *In vitro*, HAV replicates in a variety of cell lines, but rather poorly. HAV is a picornavirus, currently classified in the genus *Enterovirus*. Genome studies based on sequence homology show that it probably does not belong in this genus and undoubtedly will soon be reclassified, probably in a genus of its own. HAV causes 'infectious' or 'epidemic' hepatitis. This disease has been recognized for many centuries, and was wrongly believed to be spread by aerosols. In fact, HAV is spread by the fecal–oral route. Outbreaks are frequently associated with consumption of shellfish. Two HAV vaccines have been developed, one inactivated and one live attenuated vaccine. The availability of assays for and vaccines against HAV means that the incidence of 'infectious hepatitis' will undoubtedly decrease sharply in future.
- *Hepatitis B virus (HBV)*: HBV causes 'serum hepatitis.' This disease was distinguished clinically from 'infectious hepatitis' in the 1930s. The virus was first identified in the late 1960s. Infection often results from inoculation with human serum – blood transfusions, transplants, or passive immunization, and is common among intravenous drug abusers who share needles. However, the virus

is also transmitted sexually, by ingestion, and from mother to child (transplacenta and via breast milk) which accounts for familial clusters of infection. All blood/organ/tissue donations in developed countries are now tested for HBV and the risk of infection by this route is extremely low. HBV does not replicate in cultured cells and this has hindered investigations of the virus. The surface antigen of the virus (HBsAg) obtained from the blood of chronic carriers was originally used as a vaccine – a very risky practice! A recombinant HBsAg vaccine produced in yeast is now used which is safe and effective.

- *Hepatitis C virus (HCV)*: HCV was first identified in 1989 and is transmitted by blood/serum and sexual or familial contacts in a similar way to HBV. During the last year, the virus has been cloned and the complete sequence of several isolates determined. These studies have caused some controversy over the virus family to which HCV should be assigned. The genome consists of single-stranded (+) sense RNA of approximately 7.5 kb, and the gene organization suggests that HCV is a flavivirus. However, the 5′ non-translated region contains three or four AUG codons – more typical of picornaviruses than flaviviruses. It has been suggested that the virus should be placed in a new genus within the family *Flaviviridae*. Several diagnostic tests have been developed, based on serological and molecular techniques and the development of vaccines is under way.

- *Hepatitis delta virus (HDV)*: In 1977, a novel nuclear antigen was discovered in patients with chronic HBV infection. This was called the delta (δ) antigen following on from the three known HBV antigens. By 1980, transmission experiments in chimpanzees showed that this agent was a defective transmissible pathogen dependent on HBV for replication – similar to plant virus satellite RNAs (see Satellites and Viroids). The δ antigen is found worldwide wherever HBV infection occurs and seems to potentiate the pathogenic effects of HBV infection. Fulminant hepatitis (with a mortality rate of approximately 80%) is ten times more common in HBV/HDV co-infections than with HBV infection alone. Since HDV replication is absolutely dependent on HBV, HDV can be controled by eliminating HBV infection (see above).

- *Hepatitis E virus (HEV)*: HEV was identified in 1990 and its entire genome has now been cloned and sequenced. The virion is a 32–34 nm non-enveloped particle containing a single-stranded (+)sense RNA genome of about 7.5 kb. The genetic organization suggests HEV may be a calicivirus, though this is problematic, since we currently know so little about other caliciviruses. The normal course of HEV infection seems to be an acute but relatively benign illness (cf. HAV), except during pregnancy when it may cause 15–30% mortality. Recombinant vaccines are currently being prepared.

- *Hepatitis F virus (HFV)*: In liver samples from patients who died of fulminant hepatitis, examination by electron microscopy has revealed 60 nm ogavirus-like particles. These have provisionally been referred to as 'HFV.' This agent appears to cause a malignant form of hepatitis, including rejection and necrosis of transplanted livers. HFV is currently being investigated by molecular biological techniques but little else is known about this virus at present.

- *Hepatitis G virus (HGV)*: More of a rumor than a virus, HGV is reported to be a paramyxovirus-like agent transmissible to chimpanzees, causing syncytial giant cell hepatitis. Nothing else is known about HGV at present.

These are not the only novel viruses which have been discovered in the last few years. Will any more new viruses be discovered in the future? You bet!

Satellites and Viroids

Satellites are small RNA molecules which are absolutely dependent on the presence of another virus for multiplication. Even viruses have their own parasites! Most satellites are associated with plant viruses, but a few are associated with bacteriophages or animal viruses, for example, the dependoviruses (Parvoviridae), which are satellites of adenoviruses. Two classes of satellite molecules can be distinguished: satellite viruses, which encode their own coat proteins, and satellite RNAs (or 'virusoids'), which use the coat protein of the helper virus. Typical properties of satellites are:

- A genome of approximately 500–2000 nucleotides of single-stranded RNA.
- Unlike defective virus genomes, there is little or no nucleotide sequence similarity between the satellite and the helper virus genome.
- They cause distinct disease symptoms in plants which are not seen with the helper virus alone.
- Replication of satellites usually interferes with the replication of the helper virus (unlike most defective virus genomes).

Viroids are very small (200–400 nt), rod-like RNA molecules with a high degree of secondary structure (Figure 8.2). They have no capsid or envelope, consisting only of a single nucleic acid molecule. Viroids are associated with certain plant diseases. Unlike satellite viruses (above), viroids show no requirement for a helper virus and exist as independent entities. Unlike satellite viruses, viroids do not encode any polypeptides and are replicated by host cell RNA polymerase II. By necessity, therefore, the replication of viroids occurs in the nucleus, unlike that of satellites, many of which are replicated in the cytoplasm. The first to be discovered and best studied is potato spindle tuber viroid (PSTVd – viroid names are abbreviated as 'Vd' to distinguish them from viruses).

There is considerable sequence variation between different viroids, and this is used as an arbitrary classification to divide viroids into species. However, all viroids share a common feature, a conserved central region believed to be involved in their replication (Figure 8.3). One group of viroids is capable of forming a hammerhead structure, giving them the enzymatic properties of a ribozyme (an autocatalytic, self-cleaving RNA molecule). This activity is used to cleave the multimeric structures produced during the course of replication. Other viroids use

Figure 8.2 Structure of a viroid RNA.

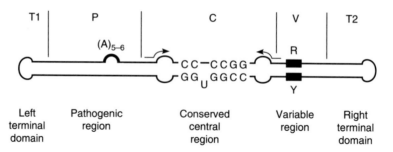

Figure 8.3 Functional regions of viroid RNA molecules.

unknown host nuclear enzymes to achieve this objective. Some viroids (e.g. cadang-cadang coconut viroid – CCCVd) cause severe and lethal disease in their host plants. Others range from no apparent pathogenic effects (e.g. hop latent viroid – HLVd) to mild disease symptoms (e.g. apple scar skin viroid – ASSVd). It is not clear how viroids cause pathogenic symptoms, but obviously these must result from some perturbation of the normal host cell metabolism. They show some similarities with certain eukaryotic host cell sequences, in particular with an intron found between the 5.8S and 25S ribosomal RNAs and with the U3 snRNA which is involved in splicing. Therefore, it has been suggested that viroids may interfere with post-transcriptional RNA processing in infected cells.

Hepatitis delta virus (HDV – see page 196) is a unique chimeric molecule with some of the properties of a satellite virus and some of a viroid (Figure 8.4; Table 8.1) which causes disease in humans. Virus preparations from HBV/HDV-infected

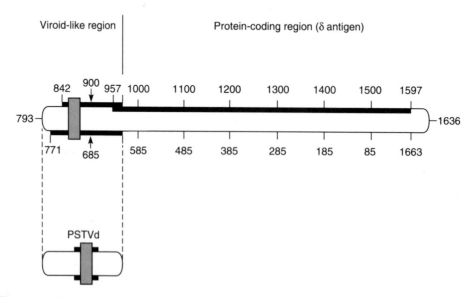

Figure 8.4 Structure of hepatitis δ virus RNA. A region at the left end of the genome strongly resembles the RNA of plant viroids such as potato spindle tuber viroid (PSTVd), which is shown for comparison.

Table 8.1 Satellite-like and viroid-like properties of hepatitis delta virus

Satellite-like properties	Viroid-like properties
Size and composition of genome: 1640 nt (about four times the size of plant viroids) Single-stranded circular RNA molecule	Sequence homology to the conserved central region involved in the replication of viroids
Dependent on HBV for replication: HDV RNA is packaged into coats consisting of lipids plus HBV-encoded proteins.	
Encodes a single polypeptide, the δ antigen	

animals contain heterologous particles distinct from those of HBV but with an irregular, ill-defined structure. These particles are composed of HBV antigens and contain the covalently closed circular HDV RNA molecule in a branched or rod-like configuration similar to that of other viroids. Unlike all other viroids, HDV encodes a protein, the δ antigen. The function of this protein is not known, but it is a nuclear phosphoprotein believed to be necessary for HDV replication.

Most viroids are transmitted by vegetative propagation, i.e. division of infected plants, although a few can be transmitted by insect vectors (non-propagative) or mechanically. Since viroids do not have the benefit of a protective capsid, viroid RNAs would be expected to be at extreme risk of degradation in the environment. However, their small size and high degree of secondary structure protects them to a large extent and they are able to persist in the environment for a sufficiently long period to be transferred to another host. HDV is transmitted by the same means as the HBV helper virus and benefits from the presence of a protective coat composed of lipid plus HBV proteins. The origins of viroids are obscure. One theory is that they may be the most primitive type of RNA genome – possibly leftovers from the 'RNA world' believed to have existed during the era of prebiotic evolution (see Chapter 3). Alternatively, they may have evolved at a much more recent time as the most extreme type of parasite. We may never know which of these alternatives is true, but viroids exist and cause disease in plants and in man.

Prions

There is a group of transmissible, chronic, progressive infections of the nervous system which were previously called 'slow virus diseases.' All of these diseases show common pathological effects and they are always fatal. There are a number of 'slow' infections of the nervous system (i.e. diseases where the period between infection and the development of clinical symptoms may be many years) which are caused by conventional viruses (e.g. measles virus, papovaviruses, herpesviruses, and retroviruses such as HTLV and HIV). However, the slow virus diseases are not caused by conventional viruses but by an altogether different type of agent.

Therefore, the term 'slow virus' is best avoided. The diseases under discussion are now known as subacute spongiform encephalopathies.

Animal infections

Scrapie

Scrapie is primarily a disease of sheep but it can also affect goats. It is the oldest known spongiform encephalopathy, first recognized several centuries ago. Scrapie occurs worldwide, but this is probably due to the movements of infected sheep which have occurred. Infected animals show severe and progressive neurological symptoms such as abnormal gait, often repeatedly scraping against fences or posts, from which behavior the disease takes its name. The incidence of the disease increases with the age of the animals. Scrapie is not seen in sheep of less than one and a half years old, which indicates that the incubation period of scrapie is at least this long. Some countries, such as Australia, have eliminated scrapie by slaughtering infected sheep and the imposition of rigorous import controls.

Transmissible mink encephalopathy (TME)

This is a scrapie-like disease of captive mink shown to be caused by feeding captive mink with scrapie-infected sheep corpses. Similarly, spongiform encephalopathy of domestic cats has recently been reported and is presumably transmitted by the same route. Therefore, it is clear that spongiform encephalopathies can be transmitted by the consumption of infected corpses (see description of kuru below).

Chronic wasting disease

This is a disease similar to scrapie which affects captive deer.

Bovine spongiform encephalopathy (BSE)

BSE was first recognized in dairy cattle in the UK in 1986 as a typical spongiform encephalopathy. Affected cattle showed altered behavior and a staggering gait, giving the disease its name in the press – 'mad cow disease.' On microscopic examination, the brains of affected cattle showed extensive spongiform degeneration. A few cases of similar disease have also been reported in captive antelopes in zoos and in cattle from a few other countries. It is believed that these cases are due to the use of contaminated foodstuffs produced in the UK – the vast majority of cases have occurred in cattle in the UK. The nutritional value of feedstuffs for domestic animals is routinely boosted by the addition of protein derived from waste meat products and bonemeal prepared from animal carcasses. In the UK in the early 1980s, the regulations controlling the rendering process were altered to permit the introduction of a less intensive, lower temperature treatment. It is believed that this change allowed the infectious agent from scrapie-infected sheep to infect cattle via contaminated foodstuffs. In 1988, new legislation was introduced to prohibit the recycling of ruminant-derived protein in ruminant foodstuffs. It is believed that this step stopped the exposure of more cattle to the disease agent. Nevertheless, the number of cases of BSE continues to rise and thousands of cases have been confirmed in each of the past few years. The median incubation period for BSE in cattle is not known accurately, but since the source of infection is presumed to have been removed in 1988, it is presumed that the incidence of BSE will begin to fall at some point in the future. However, the extent to which the numbers of reported cases will decline depends critically on two factors, first on whether the infectious agent is transmitted vertically in cattle (i.e. from cow to calf), and second, the extent to which the British dairy herd has been contaminated. Neither of these facts are known at present. Undoubtedly BSE could be controlled by a rigorous slaughter program just as scrapie has been eliminated in some countries, but the cost of this solution could be immense.

Human infections

Kuru

Kuru was the first human spongiform encephalopathy to be investigated in detail and is one of the most fascinating stories to have emerged from any epidemiological investigation. The disease occurred primarily in 169 villages occupied by the Fore tribespeople in the highlands of New Guinea. The first recorded cases were noted in the 1950s and involved progressive loss of voluntary neuronal control, followed by death less than one year after the onset of symptoms. The key to the origin of the disease was provided by the profile of its victims – it was never seen in young children, rarely in adult men and was most common in both male and female adolescents and in adult women. The Fore people are believed to have practised ritual cannibalism as a rite of mourning for their dead. Women and children participated in these ceremonies but adult men did not take part, which explains the age/sex distribution of the disease. The incubation period for kuru can be in excess of 30 years, but in most cases is somewhat shorter. The practice of ritual cannibalism was discouraged in the late 1950s and the incidence of kuru has declined dramatically over the last few decades. Kuru is now very rare and occurs only in individuals who participated in cannibalistic feasts over 30 years ago.

Creutzfeldt–Jakob disease (CJD)

Unlike kuru, CJD is a rare, apparently spontaneous form of senile dementia which occurs in numerous human populations worldwide at the rate of about one case per million per year. Occasionally there is evidence of a familial pattern of inheritance, i.e. vertical transmission. CJD is characterized by mental degeneration followed by motor dysfunction and death within one year of the onset of symptoms. The mean age at death is 57 years and the minimum age at death about 27 years. Some cases of CJD are known to be associated with transplantation of tissues from infected donors, such as corneal transplants, or the use of growth hormone derived from human pituitary glands, indicating that it has at least an infectious component. The majority of cases are impossible to explain other than by assuming the spontaneous appearance of the disease on an individual basis. (It is believed that the relatively sudden onset of kuru in New Guinea resulted from cannibalisitic consumption of one such spontaneous case).

Gerstmann–Sträussler syndrome (GSS)

This is a disease which is similar to and generally believed to be a variant form of CJD. GSS is particularly notable for the deposition of amyloid plaques in the brain (see below) and the somewhat longer time to death than in typical cases of CJD.

All these conditions show the same basic histochemical abnormalities and are therefore believed to be caused by similar mechanisms. Pathological changes include a defect in plasma membrane formation followed by vacuolation of neurons, astrocytes, and oligodendrocytes. At a later stage, the gray matter of the CNS shows extensive neuronal loss and has a 'spongiform' appearance under the microscope, i.e. it is full of holes. Intracellular cytochemical changes include the accumulation of large plaques composed of glial fibrillary acidic protein (GFAP) and fibrils consisting of another cellular protein (see below). In this respect, these transmissible diseases closely resemble the pathologic symptoms seen in Alzheimer's disease, a form of senile dementia. However, there is currently no evidence that Alzheimer's disease is transmissible and it is not believed that it is caused by an infectious agent. Rather, it would appear that Alzheimer's disease and the transmissible spongiform encephalopathies represent convergent pathologic conditions probably derived from different sources.

Initially, there was reason to suspect that these transmissible spongiform encephalopathies were caused by a virus. A similar neurodegenerative disease could be induced by inoculation of preparations made from infected tissues into the CNS of chimpanzees and filtration studies established that the size of the infectious agent responsible was comparable to that of known viruses. Research was helped by the discovery that the agent could also be transferred to rodents, although the incubation times to the onset of disease in mice were still of the order of one year and this drawback has slowed down progress significantly. However, further investigation of the characteristics of the agent indicated that it did not appear to be a virus of any known type since it displayed the following properties:

- *Resistance to heat inactivation*: Infectivity is not inactivated by heating to 90°C for 30 min. Recently, it has been reported that some infectious activity may even be retained after treatment for one hour at 360°C.
- *Resistance to radiation damage*: Infectivity was found to be resistant to short-wave ultraviolet radiation and to ionizing radiation. These treatments inactivate infectious organisms by causing damage to the genome. There is an inverse relationship between the size of target nucleic acid molecule and the dose of radioactivity or ultraviolet light needed to inactivate them, i.e. large molecules are sensitive to much lower doses than are smaller molecules (Figure 8.5). The scrapie agent was found to be highly resistant to both ultraviolet light and ionizing radiation, indicating that any nucleic acid present must be extremely small, probably less than 50 nucleotides.

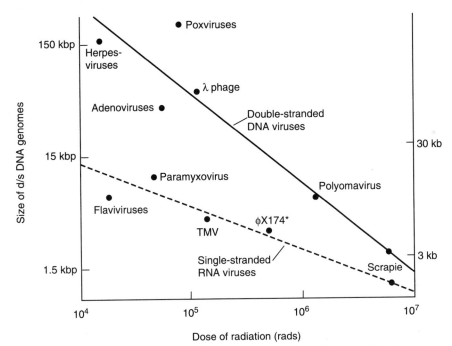

Figure 8.5 Radiation sensitivity of infectious agents, compiled from published sources. The dose of ionizing radiation required to destroy the infectivity of an infectious agent is dependent on the size of its genome. Larger genomes (e.g. double-stranded DNA viruses – upper line) present a larger 'target' and therefore are more sensitive than smaller genomes (e.g. single-stranded RNA viruses – lower line). The scrapie agent is considerably more resistant to radiation than any known virus (note log scale on vertical axis). ɸX174 has a single-strained RNA genome.

- *Resistance* to DNAse and RNAse treatment, to psoralens, and to Zn^{2+}-catalyzed hydrolysis. All these treatments inactivate nucleic acids.
- *Sensitivity* to urea, sodium dodecyl sulfate (SDS), phenol, and other protein-denaturing chemicals.

This evidence suggests that the agent is composed exclusively of protein and that any nucleic acid, if present, must be extremely small. Study of infectious material resulted in the identification of a protein of 254 amino acids which appeared to be responsible for infectivity. In recent years, extensive biochemical investigation of this protein has been carried out. To date, there is no positive evidence that any nucleic acid component is associated with this protein or is required for infectivity. In 1982, Stanley Prusiner proposed the term **prion** for this protein, which is now referred to as PrP ('prion protein'). Prusiner himself acknowledged that the idea that a protein devoid of any nucleic acid could itself be an infectious agent was heretical. Worse was to come. It was discovered that PrP is encoded by a single-copy cellular gene and that it is found in 'normal' (i.e. apparently uninfected) cells. There are two forms of PrP (Figure 8.6):

- PrP^C: This is the normal cellular form of the protein, found predominantly on

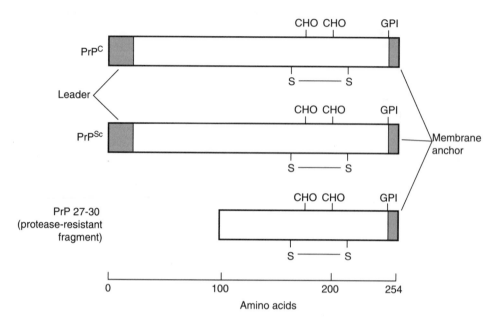

Figure 8.6 Structure of the prion protein, PrP.

the surface of neurons. PrP^C is sensitive to and completely degraded by protease digestion.

- PrP^{Sc}: This is a form of the protein found in diseased tissues. Unlike PrP^C, PRP^{Sc} is partially resistant to protease digestion, resulting in the formation of a 141 amino acid protease-resistant fragment which accumulates as fibrils in infected cells. Only a proportion of the total PrP in diseased tissue is present as PrP^{Sc}, but this has been shown to be the infectious form of the PrP protein, since extensively purified PrP^{Sc} is infectious when used to inoculate experimental animals.

Extensive studies have revealed no obvious differences (e.g. changes in amino acid sequence) between PrP^C and PrP^{Sc} which could account for the different properties of the two forms. It has therefore been suggested that functional differences must be the result of conformational changes, but this has not been proved and the three-dimensional structure of PrP is not yet known. Neither form of PrP induces the formation of antibodies or any other immunological reaction since they are both recognized by the immune system as 'self' antigens. In several human families which show a high incidence of spontaneous GSS, particular mutations in the gene encoding PrP have been observed. However, even in these cases, there is no detectable difference between PrP^C and PrP^{Sc}. This has been explained by suggesting that these mutations might facilitate the spontaneous conversion of PrP^C into PrP^{Sc}, putting individuals carrying these mutations at greatly increased risk of developing GSS. It has been suggested that PrP^{Sc} catalytically converts PrP^C into PrP^{Sc} and that this is the basis for the disease symptoms.

The prion protein is only one side of this disease equation. Transfer of the scrapie agent into mice revealed that the genetic composition of the recipient also influences the pathogenesis of spongiform encephalopathies. Genes have been identified in mice, hamsters, and sheep which influence the incubation time before the onset of disease symptoms and appear to mediate a 'species barrier' which must be overcome before animals can be efficiently infected by PrP^{Sc} from another species (Figure 8.7). (In this context, 'efficiently' refers to the length of the incubation period before disease symptoms become apparent). In the past few years, transgenic mice have been used to try to unravel the role of PrP in spongiform encephalopathy. These experiments include the construction of mice carrying PrP genes from different species (Figure 8.7). A strain of mice entirely lacking the PrP protein has been constructed by disrupting the gene which encodes the protein. Remarkably, and contrary to what may have been expected, these animals show no detectable developmental, behavioral, or biochemical abnormalities whatsoever. While this result does not answer the question of what the normal role of PrP is, it suggests that the pathogenesis of the spongiform encephalopathies might result from the accumulation of protease-resistant forms of PrP^{Sc} rather than from a deficiency in unmodified PrP^{C}.

The prion hypothesis is revolutionary and has justifiably met with a somewhat sceptical reception. Although evidence in its favor is growing, there are a few

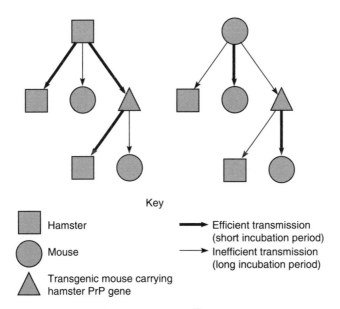

Key

Hamster   ⟶ Efficient transmission
                          (short incubation period)
Mouse   → Inefficient transmission
                          (long incubation period)

Transgenic mouse carrying
hamster PrP gene

Figure 8.7 Experimental transmission of scrapie to animals demonstrates an apparent species barrier. Hamster-to-hamster and mouse-to-mouse transmission results in the onset of disease after a relatively short incubation period. Transmission from one host species to another is much less efficient and disease only occurs after a much longer incubation period. Transmission of hamster-derived PrP to transgenic mice carrying several copies of the hamster PrP gene is much more efficient, whereas transmission of mouse-derived PrP to the transgenic mice is less efficient. Subsequent transmission from the transgenic mice implies that some modification of the properties of the agent seems to have occurred.

inconsistencies which have not been adequately explained. One of these is the existence of several strains of PrP which reproducibly show different disease phenotypes in mice (e.g. incubation period or pattern of lesions in the CNS). How could a single, small, apparently invariant protein cause so many different patterns of disease? It has been suggested that PrP^{Sc} converts PrP^{C} into the infectious form by altering its conformation and that each different strain represents a different conformation of the protein. This has not been proved or disproved since the structure of PrP is not yet known, but it seems to be stretching belief that such a small protein could assume so many different conformations. In addition, it is necessary to explain how the behavior of PrP^{Sc} apparently changes once the species barrier has been overcome (Figure 8.7).

Because of this and because of the general reluctance to accept such an outrageous concept as the prion hypothesis, an alternative theory has been put forward – the virino (or nucleoprotein) theory. The virino theory suggests that there is, in fact, a small nucleic acid molecule associated with PrP. This would not be a genomic molecule in the sense that it encodes the information necessary for the synthesis of any part of PrP, but the presence of this additional molecule is required for the disease-causing properties of PrP^{Sc} to be apparent. Nucleotide sequence variation of this nucleic acid molecule could give rise to the strain variation which has been observed in different isolates of the scrapie agent. As stated above, such an accessory molecule has never been detected, even by using such powerful techniques as the polymerase chain reaction. However, negative results cannot prove that such an entity does not exist. Therefore, the onus must be on the supporters of the virino theory to identify the missing accessory nucleic acid molecule, which to date, has not been achieved.

It is possible to construct numerous alternative theories of varying degrees of complexity (and plausibility!) to fit the experimental data, some of which go some way towards reconciling the conflict between the prion and virino theories. At the present time, informed observers can do no more than weigh the available evidence and make their own decisions about what they would like to believe. Science progresses by the construction of experimentally verifiable hypotheses. For many years, research into spongiform encephalopathies has been agonizingly slow, since each individual experiment has taken at least one and in some cases many years to complete. With the advent of molecular biology, this has now become a fast-moving and dynamic field. The next few years will undoubtedly reveal much more information about the cause of these diseases and will probably deliver much food for thought about the interaction between infectious agents and the host in the pathogenesis of infectious diseases.

Summary

A variety of novel infectious agents cause disease in plants, in animals, and in man. Some conventional viruses can be placed in this category, including both newly discovered viruses and emerging viruses which have come to prominence as a result of altered patterns of disease. In addition to these, several types of non-viral, subcellular pathogens have disease-causing potential. These include satellites, viroids, and prions. Conventional strategies to combat virus infections, such as drugs and vaccines, have no effect on these unconventional agents. A full understanding of the biology of these novel infectious entities will be necessary before means of treatment for the diseases they cause will become available.

Further Reading

Diener, T.O. (Ed.) (1987). *The Viroids*. Plenum Press, New York.

Morse, S.S. and Schluederberg, A. (1990). Emerging viruses: the evolution of viruses and viral diseases. *J. Infect. Dis.* **162**: 1–7.

Prusiner, S.B. (1991). Molecular biology of prion diseases. *Science* **252**: 1515–22.

Bibliography

Chesebro, B. (Ed.) (1991). Scrapie and other transmissible spongiform encephalopathies. *Curr. Topics Microbiol. Immunol.* **172**.

Fagan, E.A. (1992). Hepatitis A to G and beyond. *Brit. J. Hosp. Med.* **47**: 127–31.

Prusiner, S.B. (1982). Novel proteinaceous infectious particles cause scrapie. *Science* **216**: 136–44.

Weissmann, C.B. (1991). The prion's progress. *Nature* **349**: 569–71.

Classification of Subcellular Infectious Agents

The importance of virus identification has been discussed in Chapter 4. Subcellular agents present a particular problem for taxonomists. They are too small to be seen without electron microscopes but very small changes in molecular structure may give rise to agents with radically different properties. The vast majority of viruses which are known have been studied because they have pathogenic potential for humans, animals, or plants. Therefore, the disease symptoms caused by infection are one criterion used to aid classification. The physical structure of a virus particle can be determined directly (by electron microscopy) or indirectly (by biochemical or serological investigation) and is also used in classification. However, the structure and sequence of the virus genome continues to increase in importance as molecular biological analysis provides a rapid and sensitive way to detect and differentiate many diverse viruses.

In 1966, the International Committee on Nomenclature of Viruses was established and produced the first unified scheme for virus classification. In 1973, this committee expanded its objectives and renamed itself the International Committee on Taxonomy of Viruses (ICTV), which meets every four years. Rules for virus taxonomy were established, some of which include:

- Latin binomial names (e.g. *Rhabdovirus carpio*) have now been abandoned, although existing names have been retained wherever possible. No person's name should be used in nomenclature. Names should have international meaning.
- A virus name should be meaningful and should consist of as few words as possible. Serial numbers or letters are not acceptable as names.
- A virus *species* is represented by a cluster of strains from a variety of sources or a population of strains from one particular source, all of which have in common a set pattern of stable properties that separates the cluster from other clusters of strains.
- A *genus* is a group of virus species sharing common characters. Approval of a

new genus is linked to the acceptance of a type species, i.e. a species which displays the typical characteristics on which the genus is based.
- A *family* is a group of genera with common characters. Approval of a new family is linked to the acceptance of a type genus.

In general terms, groups of related viruses are divided into families whose names are capitalized, italicized, and end in the suffix '*viridae*' (e.g. *Poxviridae*). In most cases, a higher level of classification than the family has not been established, although this is now changing with the consideration of evolutionary relationships between families and the arrival of the concept of 'superfamilies' equivalent to the 'orders' of formal nomenclature (see Chapter 3). In a few cases, very large families have been subdivided into subfamilies, written as above (e.g. *Chordopoxviridae*). In most cases, families consist of a collection of genera whose names are capitalized, italicized, and end in the suffix '*virus*' (e.g. *Orthopoxvirus*). However, many virus genera are sufficiently different to all others to be recognized as a separate group in their own right and not as part of a larger family. Each genus contains a number of virus species, whose names are not capitalized or italicized (e.g. vaccinia virus). Some genera are monotypic, i.e. contain only one species. Subspecies, strains, isolates, variants, mutants, and artificially created laboratory recombinants are not officially recognized by the ICTV.

The fifth report of the ICTV was published in 1991 (*Classification and Nomenclature of Viruses*. R.I.B. Francki, C.M. Fauquet, D.L. Knudson and F. Brown (Eds), *Archives of Virology*, Supplement 2, 1991). A total of 2430 virus species belonging to 72 families or groups are recognized in this report. These well-characterized viruses are probably a small proportion of the total number of viruses which exist. By the time of the next ICTV meeting, undoubtedly many more viruses will have been added to this list.

Satellites and viroids are not officially classified by the ICTV in the same way as conventional viruses. However, criteria have been established which divide plant virus satellites into four groups:

- *Type A*: An RNA of more than 700 nt which encodes a structural capsid protein forming satellite-specific particles.
- *Type B*: An RNA of more than 700 nt which encodes a non-structural protein.
- *Type C*: A linear RNA of less than 700 nt which does not encode any proteins.
- *Type D*: A circular RNA of less than 700 nt which does not encode any proteins.

Similarly, viroids are not officially recognized by the ICTV, but are grouped on the basis of nucleotide sequence conservation of the central region involved in replication. An arbitrary level of 90% sequence similarity is the criterion used to assign viroids to groups, each based on a type member.

Table A1 Viruses

Family or group	Genera	Type member	Particle morphology	Envelope	Genome	Host
Adenoviridae	Mastadenovirus	Human adenovirus 2	Icosahedral	–	Linear d/s DNA	Vertebrates
	Aviadenovirus	Fowl adenovirus 1				
Alfalfa mosaic virus	–	–	Bacilliform	–	3 (+)strands RNA	Plants
Arenaviridae	Arenavirus	Lymphocytic choriomeningitis	Spherical	+	2 (–)strands RNA	Vertebrates
Baculoviridae	Nuclear polyhedrosis	–	Bacilliform	+	Supercoiled DNA	Invertebrates
	Granulosis virus					
Birnaviridae	Birnavirus	Infectious pancreatic necrosis	Icosahedral	–	2 segments d/s RNA	Vertebrates, invertebrates
Bromoviridae	Bromovirus	Brome mosaic	Icosahedral	–	3 (+)strands RNA	Plants
Bunyaviridae	Bunyavirus	Bunyamwera	Spherical	+	3 (–)strands RNA	Vertebrates
	Phlebovirus	Sandfly fever				
	Nairovirus	Crimean–Congo hemorrhagic fever				
		Hantaan				
	Hantavirus					
	Tospovirus	Tomato spotted wilt				Plants
Caliciviridae	Calicivirus	Vesicular exanthema of swine	Icosahedral	–	1 (+)strand RNA	Vertebrates
Capillovirus	Capillovirus	Apple stem grooving	Rod	–	1 (+)strand RNA	Plants
Carlavirus	Carlavirus	Carnation latent	Rod	–	1 (+)strand RNA	Plants
Carmovirus	Carmovirus	Carnation mottle	Isometric	–	1 (+)strand RNA	Plants
Caulimovirus	Caulimovirus	Cauliflower mosaic	Isometric	–	Circular d/s DNA	Plants
Closterovirus	Closterovirus	Beet yellows	Rod	–	1 (+)strand RNA	Plants
Commelina yellow mottle virus	–	–	Bacilliform	–	Circular d/s DNA	Plants
Comovirus	Comovirus	Cowpea mosaic	Isometric	–	1 (+)strand RNA	Plants
Coronaviridae	Coronavirus	Avian infectious bronchitis	Pleomorphic	+	1 (+)strand RNA	Vertebrates
Corticoviridae	Corticovirus	PM2 phage	Isometric	–	Supercoiled d/s DNA	Bacteria

Family	Genus	Type species	Morphology	±	Genome	Host
	Cryptovirus	White clover cryptic	Isometric	–	2 segments d/s RNA	Plants
	Cucumovirus	Cucumber mosaic	Isometric	–	3 (+)strands RNA	Plants
Cystoviridae	*Cystovirus*	φ6 phage	Isometric	+	3 segments d/s RNA	Bacteria
	Dianthovirus	Carnation ringspot	Isometric	–	2 (+)strands RNA	Plants
	Fabavirus	Broad bean wilt	Isometric	–	2 (+)strands RNA	Plants
Filoviridae	*Filovirus*	Marburg	Bacilliform	+	1 (–)strand RNA	Vertebrates
Flaviviridae	*Flavivirus*	Yellow fever	Spherical	+	1 (+)strand RNA	Vertebrates, invertebrates
	Pestivirus	Bovine viral diarrhea				
	Hepatitis C virus					
	Furovirus	Soilborne wheat mosaic	Rod	–	2 (+)strands RNA	Plants
	Geminivirus	Maize streak	Isometric	–	1–2 circular strands s/s DNA	Plants
Hepadnaviridae	*Orthohepadnavirus*	Hepatitis B	Isometric	+	Circular d/s DNA	Vertebrates
	Avihepadnavirus	Duck hepatitis				
Herpesviridae	*Simplexvirus*	Herpes simplex	Isometric	+	Linear d/s DNA	Vertebrates
	Varicellovirus	Varicella-zoster				
	Cytomegalovirus	Human cytomegalovirus				
	Muromegalovirus	Murine cytomegalovirus				
	Lymphocryptovirus	Epstein–Barr				
	Rhadinovirus	Ateline herpesvirus				
	Hordeivirus	Barley stripe mosaic	Helical	–	3 (+)strands RNA	Plants
	Ilarvirus	Tobacco streak	Isometric	–	3 (+)strands RNA	Plants
Inoviridae	*Inovirus*	Coliphage fd	Rod	–	Circular s/s DNA	Bacteria, mycoplasma
	Plectrovirus	Acholeplasma phage				
Iridoviridae	*Iridovirus*	Chilo iridescent	Icosahedral	+	Linear d/s DNA	Vertebrates, invertebrates
	Chloriridovirus	Mosquito iridescent				
	Ranavirus	Frog virus 3				
	Lymphocystisvirus	Flounder isolate				
		Goldfish virus				
Leviviridae	*Levivirus*	Coliphage MS2	Icosahedral	–	1 (+)strand RNA	Bacteria
	Allolevivirus	Coliphage Qβ				
Lipothrixviridae	*Lipothrixvirus*	*Thermoproteus* phage 1	Rod	+	Linear d/s DNA	Bacteria

Table A1 Continued

Family or group	Genera	Type member	Particle morphology	Envelope	Genome	Host
Luteovirus	*Luteovirus*	Barley yellow dwarf	Isometric	–	1 (+)strand RNA	Plants
Maize chlorotic dwarf virus		—	Isometric	–	1 (+)strand RNA	Plants
Marafivirus	*Marafivirus*	Maize rayado fino	Isometric	–	1 (+)strand RNA	Plants
Microviridae	*Microvirus*	Coliphage φX174	Icosahedral	–	Circular s/s DNA	Bacteria
	Spirovirus	*Spiroplasma* phages				
	Mac-1 phage	—				
Myoviridae	T4 phage	—	Tailed phage	–	Linear d/s DNA	Bacteria
Necrovirus	*Necrovirus*	Tobacco necrosis	Isometric	–	1 (+)strand RNA	Plants
Nepovirus	*Nepovirus*	Tobacco ringspot	Isometric	–	2 (+)strands RNA	Plants
Nodaviridae	*Nodavirus*	Nodamura	Isometric	–	2 (+)strands RNA	Invertebrates
Orthomyxoviridae	*Influenzavirus* A & B	—	Helical	+	8 (–)strands RNA	Vertebrates
	Influenzavirus C	—				
Papovaviridae	*Papillomavirus*	Rabbit papillomavirus	Icosahedral	–	Circular d/s DNA	Vertebrates
	Polyomavirus	Mouse polyomavirus				
Paramyxoviridae	*Paramyxovirus*	Newcastle disease	Helical	+	1 (–)strand RNA	Vertebrates
	Morbillivirus	Measles				
	Pneumovirus	Respiratory syncytial				
Parsnip yellow fleck virus		—	Isometric	–	1 (+)strand RNA	Plants
Partitiviridae	*Partitivirus*	*Gaeumannomyces graminis* virus	Isometric	–	2 segments d/s RNA	Fungi
Parvoviridae	*Parvovirus*	Minute virus of mice	Icosahedral	–	1 (–)strand RNA	Vertebrates, invertebrates
	Dependovirus	Adeno-associated virus				
	Densovirus	*Galleria* densovirus				
Pea enation mosaic virus		—	Isometric	–	2 (+)strands RNA	Plants

Family	Genus	Species/Example	Morphology	Envelope	Nucleic acid	Host
Phycodnaviridae	*Phycodnavirus*	*Paramecium bursaria chlorella virus 1*	Icosahedral	–	Linear d/s DNA	Algae
Picornaviridae	*Enterovirus* *Hepatovirus* *Cardiovirus* *Rhinovirus* *Aphthovirus*	Poliovirus 1 Hepatitis A Encephalomyocarditis Human rhinovirus 1A Foot-and-mouth disease	Icosahedral	–	1 (+)strand RNA	Vertebrates, invertebrates
Plasmaviridae	*Plasmavirus*	Acholesplasma phage	Pleomorphic	+	Circular d/s DNA	Bacteria
Podoviridae		T7 phage —	Tailed phage	–	Linear d/s DNA	Bacteria
Polydnaviridae	*Ichnovirus* *Brachovirus*	*Campoletis sonorensis* *Cotesia melanoscela*	Rod, fusiform	+	Supercoiled d/s DNA	Invertebrates
Potexvirus	*Potexvirus*	Potato X virus 4	Rod	–	1 (+)strand RNA	Plants
Potyvirus	*Potyvirus*	Potato virus Y	Rod	–	1 (+)strand RNA	Plants
Poxviridae	*Orthopoxvirus* *Parapoxvirus* *Avipoxvirus* *Capripoxvirus* *Leporipoxvirus* *Suipoxvirus* *Molluscipoxvirus* *Yatapoxvirus* *Entomopoxvirus A/B/C*	Vaccinia Orf Fowlpox Sheep pox Myxoma Swinepox Molluscum contagiosum Yaba monkey tumor (Insect poxviruses)	Ovoid	+	Linear d/s DNA	Vertebrates, invertebrates
Reoviridae	*Orthoreovirus* *Orbivirus* *Coltivirus* *Rotavirus* *Aquareovirus* *Cypovirus* *Plant reovirus 1/3* *Fijivirus*	Reovirus 1 Bluetongue Colorado tick fever Human rotavirus Golden shiner Cytoplasmic polyhedrosis Wound tumor Fiji disease	Icosahedral	–	10–12 segments d/s RNA	Vertebrrates, invertebrates, plants
Retroviridae	B-type C-type D-type	Mouse mammary tumor Murine leukemia Mason–Pfizer monkey virus	Spherical	+	Dimer (+)strand RNA	Vertebrates

Table A1 Continued

Family or group	Genera	Type member	Particle morphology	Envelope	Genome	Host
	HTLV–BLV group	Human T-cell leukemia				
	Spumavirus	Human foamy				
	Lentivirus	Human immunodeficiency				
Rhabdoviridae	*Vesiculovirus*	Vesicular stomatitis	Bacilliform	+	1(−)strand RNA	Vertebrates, invertebrates, plants
	Lyssavirus	Rabies				
	Plant rhabdovirus A/B	Lettuce necrotic yellows				
Siphoviridae	λ phage group	—	Tailed phage	−	Linear d/s DNA	Bacteria
Sobemovirus	*Sobemovirus*	Southern bean mosaic	Icosahedral	−	1 (+)strand RNA	Plants
Sulfolobus shibatae virus (SSV-1)		—	Lemon-shape	+	Circular d/s DNA	Bacteria
Tectiviridae	*Tectivirus*	Phage PRD1	Icosahedral	−	Linear d/s DNA	Bacteria
Tenuivirus	*Tenuivirus*	Rice stripe	Rod	−	4 (−?)strands RNA	Plants
Tetraviridae	*Tetravirus*	*Nudaurella* β virus	Icosahedral	−	1 (+)strand RNA	Invertebrates
Tobamovirus	*Tobamovirus*	Tobacco mosaic	Rod	−	1 (+)strand RNA	Plants
Tobraviridae	*Tobravirus*	Tobacco rattle	Rod	−	2 (+)strands RNA	Plants
Togaviridae	*Alphavirus*	Sindbis	Spherical	+	1 (+)strand RNA	Vertebrates, invertebrates
	Rubivirus	Rubella				
	Arterivirus	Equine arteritis				
Tombusvirus	*Tombusvirus*	Tomato bushy stunt	Isometric	−	1 (+)strand RNA	Plants
Totiviridae	*Totivirus*	*Saccharomyces cerevisiae* virus L-A	Isometric	−	1 segment d/s RNA	Fungi
Tymovirus	*Tymovirus*	Turnip yellow mosaic	Icosahedral	−	1 (+)strand RNA	Plants

Table A2 Satellites

Helper virus	Helper virus group	Six (kb)	Type
Arabis mosaic	*Nepovirus*	0.3	D
Arabis mosaic	*Nepovirus*	1.1	B
Artichoke mottled crinkle	*Tombusvirus*	0.7	C (?)
Barley yellow dwarf	*Luteovirus*	0.32	D
Beet western yellows	*Luteovirus*	3.1	B
Carnation Italian ringspot	*Tombusvirus*	0.7	C (?)
Chicory yellow mottle	*Nepovirus*	0.46	D
Chicory yellow mottle	*Nepovirus*	1.1	B
Cucumber mosaic	*Cucumovirus*	0.3	C
Cymbidium ringspot	*Tombusvirus*	0.7	D
Grapevine Bulgarian latent	*Nepovirus*	*c.* 1.7	B
Grapevine fanleaf	*Nepovirus*	1.1	B
Groundnut rosette	(?)	0.9	B (?)
Lucerne transient streak	*Sobemovirus*	0.32	D
Maize white line mosaic	(?)	*c.* 1.3	A
Myrobalan latent ringspot	*Nepovirus*	*c.* 1.2	B
Panicum mosaic	(?)	0.4	C
Panicum mosaic	(?)	0.8	A
Pea enation mosaic	—	*c.* 0.8	B (?)
Peanut stunt	*Cucumovirus*	0.4	C
Pelargonium leaf curl	*Tombusvirus*	0.7	C (?)
Petunia asteroid mosaic	*Tombusvirus*	0.7	C (?)
Solanum nodiflorum mottle	*Sobemovirus*	0.38	D
Strawberry latent ringspot	*Nepovirus*	*c.* 1.2	B
Subterranean clover mottle	*Sobemovirus*	0.33 + 0.39	D
Tobacco mosaic	*Tobamovirus*	1.1	A
Tobacco necrosis	*Necrovirus*	0.62	C
Tobacco necrosis	*Necrovirus*	1.2	A
Tobacco ringspot	*Nepovirus*	0.3	D
Tomato black ring	*Nepovirus*	1.4	B
Tomato bushy stunt	*Tombusvirus*	0.7	C
Turnip crinkle	*Carmovirus*	0.2–0.3	C
Velvet tobacco mottle	*Sobemovirus*	0.37	D

Table A3 Viroids

Viroid	Abbreviation	Size (nt)	Group
Apple scar skin	ASSVd	330	ASSVd
Australian grapevine	AGVd	369	ASSVd
Grapevine yellow speckle 1	GYSVd 1	367	ASSVd
Grapevine yellow speckle 2	GVSVd 2+	363	ASSVd
Avocado sunblotch	ASBVd	247	ASBVd
Burdock stunt	BSVd	?	?
Chrysanthemum stunt	CSVd	356	PSTVd
Citrus exocortis	CEVd	370–375	PSTVd
Coconut Cadang-Cadang	CCCVd	246	PSTVd
Coconut tinangaja	CTiVd	254	PSTVd
Coleus blumei	CbVdl	?	?
	CbVd2	?	?
	CbVd3	248	
Columnea latent	CLVd	370	PSTVd
Hop latent	HLVd	256	PSTVd
Hop stunt	HSVd	297–303	PSTVd
Peach latent mosaic	PLMVd	?	?
Potato spindle tuber	PSTVd	359	PSTVd
Tomato apical stunt	TASVd	360	PSTVd
Tomato bunchy top	TBTVd	?	?
Tomato plant macho	TPMVd	360	PSTVd

Appendix 2

Glossary and Abbreviations

Terms shown in the text in **bold** print are defined in this glossary.

abortive infection   The initiation of infection without completion of the infectious cycle and, therefore, without the production of infectious particles (cf. **productive infection**).

adjuvant   A substance included in a medication to improve the action of the other constituents; usually, a component of vaccines which boosts their immunogenicity, e.g. aluminum sulfate.

ambisense   A single-stranded virus genome which contains genetic information encoded in both the positive (i.e. virus-sense) and negative (i.e. complementary) orientations, e.g. *Bunyaviridae* and *Arenaviridae* (see Chapter 3).

apoptosis   The genetically programmed death of certain cells which occurs during various stages in the development of multicellular organisms and may also be involved in control of the immune response.

assembly   A late phase of viral replication during which all the components necessary for the formation of a mature virion collect at a particular site in the cell and the basic structure of the virus particle is formed (see Chapter 4).

attachment   The initial interaction between a virus particle and a cellular receptor molecule; the phase of viral replication during which this occurs (see Chapter 4).

attenuated   A pathogenic agent which has been genetically altered and displays decreased virulence. Attenuated viruses are the basis of live virus vaccines (see Chapter 6).

autocrine   The production by a cell of a growth factor which is required for its own growth. Such positive feedback mechanisms may result in cellular transformation (see Chapter 7).

avirulent
An infectious agent which has *no* disease-causing potential. It is doubtful if such agents really exist – even the most innocuous organisms may cause disease in certain circumstances, e.g. in immunocompromised hosts.

bacteriophage
A virus which replicates in a bacterial host cell.

bp
Base pair – a single pair of nucleotide residues in a double-stranded nucleic acid molecule held together by Watson–Crick hydrogen bonds (see **kbp**).

budding
A mechanism involving release of virus particle from an infected cell by extrusion through a membrane. The site of budding may be at the surface of the cell or may involve the cytoplasmic or nuclear membranes depending on the site of assembly. Virus **envelopes** are acquired during budding.

capsid
The protective protein coat of a virus particle (see Chapter 2).

chromatin
The ordered complex of DNA plus proteins (histones and non-histone chromosomal proteins) found in the nucleus of **eukaryotic** cells.

***cis*-acting**
A genetic element which affects the activity of contiguous (i.e. on the same nucleic acid molecule) genetic regions; e.g. transcriptional promoters and enhancers are *cis*-acting sequences adjacent to the genes whose transcription they control.

complementation
The interaction of virus gene products in infected cells that results in the yield of one or both of the parental mutants being enhanced while their genotypes remain unchanged.

cytopathic effect (c.p.e.)
Cellular injury caused by virus infection; the effects of virus infection on cultured cells, visible by microscopic or direct visual examination (see Chapter 7).

defective interfering (D.I.) particles
Particles encoded by genetically deleted virus genomes which lack one or more essential functions for replication.

d/s
Double-stranded (nucleic acid).

eclipse period
An early phase of infection when virus particles have broken down after penetrating cells, releasing their genomes within the host cell as a prerequisite to replication. Often used to refer specifically to bacteriophages (see Chapter 4).

emergent virus
A virus identified as the cause of an increasing incidence of disease, possibly as a result of changed environmental or social factors (see Chapter 8).

endemic	A pattern of disease which recurs or is commonly found in a particular geographic area (cf. **epidemic**).
enhancers	*Cis*-acting genetic elements which potentiate the transcription of genes or translation of mRNAs.
envelope	An outer (bounding) lipoprotein bilayer membrane possessed by many viruses. (NB Some viruses contain lipid as part of a complex outer layer, but these are not usually regarded as enveloped unless a bilayer unit membrane structure is clearly demonstrable).
epidemic	A pattern of disease characterized by a rapid increase in the number of cases occurring and widespread geographical distribution (cf. **endemic**). An epidemic which encompasses the entire world is known as a **pandemic**.
eukaryote	An organism whose genetic material is separated from the cytoplasm by a nuclear membrane and divided into discrete chromosomes.
exon	A region of a gene expressed as protein after the removal of **introns** by post-transcriptional splicing.
fusion protein	A virus protein required and responsible for fusion of the virus **envelope** (or sometimes, the **capsid**) with a cellular membrane and consequently, for entry into the cell (see Chapter 4).
genome	The nucleic acid comprising the entire genetic information of an organism.
helix	A cylindrical solid formed by stacking repeated subunits in a constant relationship with respect to their amplitude and pitch (see Chapter 2).
hemagglutination	The (specific) agglutination of red blood cells by a virus or other protein.
heterozygosis	Aberrant packaging of multiple genomes may on occasion result in multiploid particles (i.e. containing more than a single genome) which are therefore heterozygous.
hnRNA	'Heterogeneous nuclear RNA' or 'heavy nuclear RNA' – the primary, unspliced transcripts found in the nucleus of eukaryotic cells.
hyperplasia	Excessive cell division or the growth of abnormally large cells. In plants, results in the production of swollen or distorted areas due to the effects of plant viruses.
hypoplasia	Localized retardation of cell growth. Numerous plant viruses cause this effect, frequently leading to **mosaicism** (the appearance of thinner, yellow areas on the leaves).

icosahedron	A solid shape consisting of 20 triangular faces arranged around the surface of a sphere – the basic symmetry of many virus particles (see Chapter 2).
immortalized cell	A cell capable of indefinite growth (i.e. number of cell divisions) in culture. On rare occasions, immortalized cells arise spontaneously but are more commonly caused by mutagenesis as a result of virus **transformation** (see Chapter 7).
inclusion bodies	Subcellular structures formed as a result of virus infection. Often a site of virus assembly (see Chapter 4).
intron	A region of a gene removed after transcription by splicing and consequently not expressed as protein (cf. **exon**).
isometric	A solid displaying cubic symmetry, of which the icosahedron is one form.
kb	1000 nucleotide residues – a unit of measurement of single-stranded nucleic acid molecules. Sometimes (wrongly) used to mean **kbp** (below).
kbp	1000 base pairs (see above) – a unit of measurement of double-stranded nucleic acid molecules.
latent period	The time after infection before the first new extracellular virus particles appear (see Chapter 4).
lysogeny	Persistent, latent infection of bacteria by **temperate bacteriophages** such as phage λ.
lytic virus	Any virus (or virus infection) which results in the death of infected cells and their physical breakdown.
maturation	A late phase of virus infection during which newly formed virus particles become infectious. Usually involves structural changes in the particle resulting from specific cleavages of capsid proteins to form the mature products, or conformational changes in proteins during assembly (see Chapter 4).
monocistronic	A messenger RNA that consists of the transcript of a single gene and which therefore encodes a single polypeptide; a virus genome which produces such an mRNA (cf. **polycistronic**).
monolayer	A flat, contiguous sheet of adherent cells attached to the solid surface of a culture vessel.
mosaicism	The appearance of thinner, yellow areas on the leaves of plants caused by the cytopathic effects of plant viruses.
mRNA	Messenger RNA.
multiplicity of infection (m.o.i.)	The (average) number of virus particles which infect each cell in an experiment.

necrosis	Cell death, particularly that caused by an external influence (cf. **apoptosis**).
negative-sense	The nucleic acid strand with a base sequence complementary to the strand which contains the protein-coding sequence of nucleotide triplets *or* a virus whose genome consists of a negative-sense strand. (Also 'minus-sense' or '(−)sense.')
non-propagative	A term describing the transmission via secondary hosts (such as arthropods) of viruses which do not replicate in the vector organism, e.g. geminiviruses.
nt	A single nucleotide residue in a nucleic acid molecule.
nucleocapsid	An ordered complex of proteins plus the nucleic acid genome of a virus.
oncogene	A gene which encodes a protein capable of inducing cellular **transformation**.
ORF	Open reading frame – a region of a gene or mRNA which encodes a polypeptide, bounded by an AUG translation start codon at the 5′ end and a termination codon at the 3′ end. Not to be confused with the poxvirus called orf.
packaging signal	A region of a virus genome with a particular nucleotide sequence or structure which specifically interacts with a virus protein(s) resulting in the incorporation of the genome into a virus particle.
pandemic	An **epidemic** which encompasses the entire world.
penetration	The phase of virus replication at which the virus particle or genome enters the host cell (see Chapter 4).
phage	See **bacteriophage**.
phenotypic mixing	Individual progeny viruses from a mixed infection which contain structural proteins derived from both parental viruses.
plaque	A localized region in a cell sheet or overlay in which cells have been destroyed or their growth retarded by virus infection.
plaque-forming units (p.f.u.)	A measure of the amount of viable virus present in a virus preparation. Includes both free virus particles and infected cells containing infectious particles ('infected centers').
plasmid	An extrachromosomal genetic element capable of autonomous replication.
polycistronic	A messenger RNA which encodes more than one polypeptide (cf. **monocistronic**).

polyprotein   A large protein which is post-transcriptionally cleaved by proteases to form a series of smaller proteins with differing functions.

positive-sense   The nucleic acid strand with a base sequence which contains the protein-coding sequence of nucleotide triplets *or* a virus whose genome consists of a positive-sense strand. (Also 'plus-sense' or '(+)sense.')

primary cell   A cultured cell explanted from an organism which is capable of only a limited number of divisions (cf. **immortalized cell**).

prion   A proteinaceous infectious particle, believed to be responsible for transmissible spongiform encephalopathies such as Creutzfeldt–Jakob disease (CJD) or bovine spongiform encephalopathy (BSE) (see Chapter 8).

productive infection A 'complete' virus infection in which further infectious particles are produced (cf. **abortive infection**).

prokaryote   An organism whose genetic material is not separated from the cytoplasm of the cell by a nuclear membrane.

promoter   A **cis-acting** regulatory region upstream of the coding region of a gene which promotes transcription by facilitating the assembly of proteins in transcriptional complexes.

propagative transmission   A term describing the transmission via secondary hosts (such as arthropods) of viruses which are able to replicate in both the primary host and the vector responsible for their transmission, e.g. plant reoviruses.

prophage   The lysogenic form of a temperate bacteriophage genome integrated into the genome of the host bacterium.

provirus   The double-stranded DNA form of a retrovirus genome integrated into the **chromatin** of the host cell.

pseudorevertant   A virus with an apparently wild type phenotype but which still contains a mutant genome – may be the result of genetic **suppression**.

quasi-equivalence   A principle describing a means of forming a regular solid from irregularly shaped subunits in which subunits in *nearly* the same local environment form *nearly* equivalent bonds with their neighbors (see Chapter 2).

quasispecies   A complex mixture of rapidly evolving and competing molecular variants of RNA virus genomes which occurs in most populations of RNA viruses.

receptor   A specific molecule on the surface of a cell to which a virus attaches as a preliminary to entering the cell. May consist of proteins or the sugar residues present on glycoproteins or glycolipids in the cell membrane (see Chapter 4).

recombination
The physical interaction of virus genomes in a **superinfected** cell resulting in progeny genomes which contain information in non-parental combinations.

release
A late phase of virus infection during which newly formed virus particles leave the cell (see Chapter 4).

replicase
An enzyme responsible for replication of RNA virus genomes (see **transcriptase**).

replicon
A nucleic acid molecule containing the information necessary for its own replication; includes both **genomes** and other molecules such as **plasmids** and **satellites**.

retrotransposon
A transposable genetic element closely resembling a retrovirus genome, bounded by long terminal repeats (see Chapter 3).

satellites
Small RNA molecules which are dependent on the presence of a helper virus for replication *but* unlike defective viruses, show no sequence homology to the helper virus genome.

shutoff
A sudden and dramatic cessation of most host cell macromolecular synthesis which occurs during some virus infections, resulting in cell damage and/or death (see Chapter 7).

splicing
Post-transcriptional modification of primary RNA transcripts which occurs in the nucleus of **eukaryotic cells** during which **introns** are removed and **exons** are joined together to produce cytoplasmic **mRNAs**.

superinfection
Infection of a single cell by more than one virus particle, especially two viruses of distinct types.

suppression
The inhibition of a mtuant phenotype by a second suppressor mutation, which may be either in the virus genome or in that of the host cell (see Chapter 3).

syncytium
A mass of cytoplasm containing several separate nuclei enclosed in a continuous membrane resulting from the fusion of individual cells.

systemic infection
An infection involving multiple parts of a multicellular organism.

temperate bacteriophage
A bacteriophage capable of establishing a **lysogenic** infection (cf. **virulent bacteriophage**).

terminal redundancy
Repeated sequences present at the ends of a nucleic acid molecule.

titer
A relative measure of the amount of virus present in any preparation.

***trans*-acting**	A genetic element encoding a diffusible product which acts on regulatory sites whether or not these are contiguous with the site from which they are produced, e.g. proteins which bind to specific sequences present on any stretch of nucleic acid present in a cell, such as transcription factors (cf. ***cis*-acting**).
transcriptase	An enzyme, usually packaged into virus particles, responsible for the transcription of RNA virus genomes (see **replicase**).
transfection	Infection of cells mediated by the introduction of nucleic acid rather than by virus particles.
transformation	Any change in the morphological, biochemical, or growth parameters of a cell.
transgenic	A genetically manipulated eukaryotic organism (animal or plant) which contains additional genetic information from another species. The additional genes may be carried and/or expressed only in the somatic cells of the transgenic organism, or in the cells of the germ line, in which case they may be inheritable by any offspring.
transposons	Specific DNA sequences which are able to move from one position in the genome of an organism to another (see Chapter 3).
triangulation number	A numerical factor which defines the symmetry of an icosahedral solid (see Chapter 2).
tropism	The types of tissues or host cells in which a virus is able to replicate.
uncoating	A general term for the events which occur after the penetration of a host cell by a virus particle, in which the virus capsid is completely or partially removed and the genome exposed, usually in the form of a nucleoprotein complex (see Chapter 4).
vaccine	A preparation containing an antigenic molecule or mixture of such molecules designed to elicit an immune response. Virus vaccines can be divided into three basic types: subunit, inactivated, and live vaccines (see Chapter 6).
vaccination	The administration of a **vaccine**.
variolation	The ancient practice of inoculating immunologically naive individuals with material obtained from smallpox patients – a primitive form of vaccination (see Chapter 1).
virino theory	An alternative to the **prion** theory which suggests that there is a small nucleic acid molecule associated with PrP (see Chapter 8).
virion	Morphologically complete (mature) infectious virus particle.

viroid	Small, infectious rod-like RNA molecules which cause diseases in plants.
virulent bacteriophage	A bacteriophage which is not capable of establishing a lysogenic infection and always kills the bacteria in which it replicates (cf. **temperate bacteriophage**).
virusoids	See **satellites**.
virus-attachment protein	A virus protein responsible for the interaction of a virus particle with a specific cellular receptor molecule.

Index